Getting Started with Scientific WorkPlace®, Scientific Word®, and Scientific Notebook®

Version 3.5

Getting Started with Scientific WorkPlace®, Scientific Word®, and Scientific Notebook®

Version 3.5

Roger Hunter
Susan Bagby
MacKichan Software, Inc.

Printed in the United States of America

10 9 8 7 6 5 4 3 2 1

Trademarks

Scientific Word, Scientific WorkPlace, Scientific Notebook, and EasyMath are registered trademarks of MacKichan Software, Inc. EasyMath is the sophisticated parsing and translating system included in *Scientific Word, Scientific WorkPlace,* and *Scientific Notebook* that allows the user to work in standard mathematical notation, request computations from the underlying computational system (Maple and/or MuPAD in this version) based on the implied commands embedded in the mathematical syntax or via menu, and receive the response in typeset standard notation or graphic form in the current document. Maple is a registered trademark of Waterloo Maple Inc. MuPAD is a trademark of SciFace GmbH. TEX is a trademark of the American Mathematical Society. TrueTEX is a registered trademark of Richard J. Kinch. Windows is a registered trademark of Microsoft Corporation. All other brand and product names are trademarks of their respective companies. The spelling portion of this product is based on Proximity Linguistic Technology. Words are checked against one or more of the following Proximity Linguibase® products:

Linguibase Name	Publisher	Number of Words	Proximity Copyright
American English	Merriam-Webster, Inc.	144,000	1997
British English	William Collins Sons & Co. Ltd.	80,000	1997
Catalan	Lluis de Yzaguirre i Maura	484,000	1993
Danish	IDE a.s	169,000	1990
Dutch	Van Dale Lexicografie bv	223,000	1996
Finnish	IDE a.s	191,000	1991
French	Hatchette	288,909	1997
French Canadian	Hatchette	288,909	1997
German	Bertelsmann Lexikon Verlag	500,000	1999
German (Swiss)	Bertelsmann Lexikon Verlag	500,000	1999
Italian	William Collins Sons & Co. Ltd.	185,000	1997
Norwegian (Bokmal)	IDE a.s	150,000	1990
Norwegian (Nynorsk)	IDE a.s	145,000	1992
Polish	MorphoLogic, Inc.		1997
Portuguese (Brazilian)	William Collins Sons & Co. Ltd.	210,000	1990
Portuguese (Continental)	William Collins Sons & Co. Ltd.	218,000	1990
Russian	Russicon		1997
Spanish	William Collins Sons & Co. Ltd.	215,000	1997
Swedish	IDE a.s	900,000	1990

This document was produced with *Scientific WorkPlace.*

Authors: *Susan Bagby* and *Roger Hunter*
Editorial Assistant: *John MacKendrick*
Manuscript Editor: *Janelle Frazer*
Compositor: *MacKichan Software, Inc.*

Contents

1 Tools for Scientific Creativity

Scientific WorkPlace, Scientific Word, Scientific Notebook, and *Scientific Viewer* set the stage for your creativity with their straightforward approach to working with mathematics and text. Together, they combine the ease of entering text and mathematics in natural notation with the power of symbolic computation, the flexibility and beauty of printed or typeset output, and the convenience of direct Internet access. Individually, they offer capabilities and features combined to meet a range of user needs.

Explore these exciting features and capabilities now. Make sure you have what you need to run the software, then complete the installation and enter the world of scientific creativity.

Understanding the Product Differences

With all four products, which we refer to together as *SW,* you can work with an easy-to-learn, easy-to-use scientific word processor. Your text appears on the same screen as mathematics, which you create using familiar mathematical notation instead of special codes. With *Scientific WorkPlace* and *Scientific Notebook,* you can perform a wide range of mathematical computations using a fully integrated computational engine, and you can create interactive course materials using the Exam Builder. Using any of the products, you can format documents by selecting a style and applying tags, rather than by entering detailed formatting instructions for each text element. *Scientific WorkPlace* and *Scientific Word* have the added capabilities of TeX, with which you can produce beautifully typeset text and mathematics that adheres to internationally accepted standards of mathematics formatting. With all four products, you can access the Internet directly. The *Viewer* version of *Scientific Notebook* provides an easy way to explore many of these features. You can use the *Viewer* to view and print documents created with the software, whether those documents are available on your local system or on the Internet.

Capabilities and Features	*WorkPlace*	*Word*	*Notebook*	*Viewer*
Document creation and printing	•	•	•	
Computation and plotting	•		•	
Typesetting with LaTeX	•	•		
Internet browsing and printing	•	•	•	•
Creation of interactive course materials with Exam Builder	•		•	

SW is characterized by a rich interface, beautiful output, natural entry of text and mathematics, and easy creation of complex documents. Version 3.5 brings even more capabilities and features to the workplace:

- **Produce your document with or without typesetting** in *Scientific WorkPlace* or *Scientific Word.* When you need the beautiful document formatting that LaTeX provides, typeset your document with the Compile, Preview, and Print commands on the Typeset menu. Typesetting provides hyphenation, kerning, ligatures, and many other precise typesetting features. Typesetting also involves the automatic generation of document elements such as footnotes, margin notes, tables of contents, and indexes. When fine formatting doesn't matter and you don't need automatically generated document elements, print your document with the Preview and Print commands on the File menu. The software prints using the Page Setup specifications and the same routines it uses to display the document in the program window. Regardless of how you produce your document, *SW* saves it as a LaTeX file. Otherwise, the two production processes are different and so are their printed results.
- **Format documents more easily.** In Version 3.5, you can use the Tag Appearance dialog box to change the screen and print appearance—but not the typeset appearance—of tagged document elements such as line spacing, font size, and justification. The properties shown in the dialog box, which are stored in a `.cst` file, affect the appearance of the document when you display it in the program window and when you use the Preview or Print commands from the File menu. *SW* ignores these settings if you typeset your document.
- **Produce portable LaTeX output.** This version of *Scientific WorkPlace* and *Scientific Word* includes a Portable LaTeX output filter that makes your documents much more portable to other LaTeX installations than documents created with earlier versions of the software. When you save a file with the Portable LaTeX filter, the line `\input{tcilatex}` is not inserted in your documents. It includes only those LaTeX packages, such as amsmath and graphicx, that should be standard on most LaTeX installations.
- **Produce TeX files for Internet use.** If you have Internet access, *SW* 3.5 can open its own `.tex` files over the World Wide Web and can launch your browser to open the file at any URL address. When you save the files locally, you can use the information they contain in your own documents. You can also perform computations on the mathematics they contain. You can open Web files created in other formats as internal (`.tex`) format.
- **Create graphic images of text and mathematics.** Now you can duplicate *SW* text and mathematics in a different location, such as on the Internet, by creating a graphic image of the information just as you see it in the document window. The careful formatting of your information is maintained when you paste a picture of it to the clipboard in Windows Enhanced Metafile format or export a picture of it to a file created in one of several widely used graphics formats.
- **Navigate easily with linked documents and new toolbars.** Document links, a special form of hypertext links, provide the mechanism for uniform navigation of related documents. You can use document links to connect online documents in a logical structure without having to include hypertext links in the actual text of your documents. Toolbars make navigation easier. Use the Link toolbar and the History toolbar as external navigation tools to speed moving around in other documents. Use the Navigate toolbar as an internal navigation tool to speed moving around in the active document.

- **Compute with Maple or MuPAD.** Version 3.5 of *Scientific WorkPlace* and *Scientific Notebook* supports two computational engines, Maple V Release 5.1 and MuPAD 1.5. Most computational features are available in both engines, although each engine can perform some computations that the other cannot. Exam Builder quizzes can run with either Maple or MuPAD. Choose the engine you want at the time of purchase.
- **Work with improved computational capabilities.** In Version 3.5, the Define Maple Name and Define MuPAD Name dialog boxes have been simplified. Computational improvements include the ability to perform Fourier and Laplace transforms. Other improvements include separate computation spaces for each open document, expanded control of random matrices, the use of single-character math names, and a Stop button to halt computations. Plotting has been speeded and is also improved with the ability to capture plot coordinates using the mouse, better plot labeling, improved plots with discontinuities, greater plot precision, and finer lines.
- **Use formula objects for dynamic calculations.** In *Scientific WorkPlace* and *Scientific Notebook,* the new formula object provides a way to enter a mathematical expression together with a computational operation. The value you see on the screen is the result of the operation.
- **Compute with physical units.** *Scientific WorkPlace* and *Scientific Notebook* provide for computations on equations containing named physical units. Whether your problem is stated in feet, meters, kilograms, or seconds, the software provides the correct unit for answers and can even convert among units.
- **Create online tests.** Version 3.5 includes a new version of the Exam Builder, which you can use to construct online multiple-choice exams, quizzes, and other algorithmically-generated course materials from your *Scientific WorkPlace* and *Scientific Notebook* source files. Exam Builder quizzes can run using either computational engine. The software takes advantage of formula objects, several special commands for generating random numbers, and the equivalents of several HTML input objects including radio buttons and check boxes. Online exams and homework can be graded automatically by the computer, so students receive instant feedback and solutions. More information about using the Exam Builder to create interactive course materials appears in the Help system.
- **Import data from your calculator.** You can import a list, vector, matrix, expression, or number to a Version 3.5 document from your Casio, Texas Instruments, or Hewlett Packard graphing calculator. Once imported, the data can be used in computations like any other data created with *Scientific WorkPlace* or *Scientific Notebook.*
- **Use new tags to create different types of notes.** Tags for several different note types have been added to the `.cst` files supplied with the system. Also, you can create your own classes of popup notes by making entries in the appropriate `.cst` files from an ASCII editor. The Body Math tag makes a very useful environment for performing scratchpad computations anywhere within a document.
- **Use improved fonts.** *SW* incorporates a new implementation of internal text storage using Unicode and associated extended Unicode TrueType fonts. The software uses its own font for mathematics, called *tciuni,* and has two new Times font packages:

Times and Mathtime. The Mathtime font package also includes mathematics set in Times. Both packages yield ligatures and improved kerning in Times text when documents are typeset, and both use the widely accepted PostScript New Font Selection Scheme for LaTeX. Files that use these packages are not only better looking but also more portable than those produced in earlier versions of the software.

- **Take advantage of new file management and file editing features.** Version 3.5 simplifies returning to read-only documents, maintaining View settings and percentages for documents, and saving and restoring user preferences. Several features make editing documents easier and more intuitive: a Context menu defines operations available for selected items; the default action for the spacebar, ENTER key, and TAB key is set to enter additional space; the empty bracket displays as a red dotted line; and the Status area gives information about linked objects and commands.
- **Work with an updated version of TeX.** Version 3.5 uses the December 1999 version of LaTeX and the latest version of the TrueTeX Formatter and Previewer. Additionally, Version 3.5 includes the Standard LaTeX format as well as the Multilingual LaTeX format.
- **Use an expanded Help system.** An expanded Help system includes typesetting tips and extensive information about performing mathematical computations.

Understanding the SW Approach

The most important feature of *SW* and the key to our software approach is the separation of content and appearance. The content of your work results from the creative process of writing—forming ideas and putting them into words. The appearance of your work results from the mechanical process of formatting—displaying the document on the screen and on the printed page in the most readable way.

The *SW* approach, which is known as *logical design,* separates the creative process of writing from the mechanical process of formatting. Logical design frees you to focus on the content instead of its format. It results in increased productivity and a more consistent, higher-quality document appearance. Logical design is different from the approach used by many other word processors. That approach, known as *visual design* (or sometimes WYSIWYG, for What You See Is What You Get), focuses on making the screen look as much like the printed page as possible. If you've used only visual systems before now, you may at first be surprised by the differences between the two approaches.

One major difference is in document formatting. When you use a visual system, you must constantly apply commands to affect the appearance of the content. For example, you might determine the appearance of a section heading by selecting the text and then choosing a font, a font size, or a typeface, or you might determine the justification of a paragraph by selecting the paragraph and then choosing centering, left justification, or right justification.

When you use a logical system, you apply commands that determine the logical structure of the content rather than define its appearance. So, to create a section heading, you select the text and apply a section heading *tag.* The format and alignment of the heading are determined by the properties of the tag. In *Scientific Notebook,* tag properties are determined by the style, a collection of commands that define the way the document appears on the screen and when you produce it without LaTeX typesetting. In *Scientific WorkPlace* and *Scientific Word*, tag properties are determined in two ways: by the style and by the document's typesetting specifications—a collection of commands that define the way the document appears when you produce it with LaTeX typesetting.

Another difference between visual and logical systems is in the display of page divisions. On the screen, visual systems divide documents into pages according to their anticipated appearance in print. To see an entire line, you often have to scroll horizontally because the screen dimensions and page dimensions do not match. In a logical system, working with pages is unnecessary, because the division of a document into pages has no connection to the document's logical structure. Thus, on the screen *SW* breaks lines to fit the window. If you resize the window, the text is reshaped to fit it so that scrolling horizontally is unnecessary. Page divisions are displayed when you preview the document.

Separating the processes of creating and formatting a document combines the best of the online and print worlds. You do the work of creating a good document; *SW* does the work of creating a beautiful one.

Using This Booklet

This brief guide to Version 3.5 of *SW* explains how to install the software on your personal computer or network. It describes how to open, close, save, and manage documents on your local system, and how to connect to the wider world of information available on the Internet. The booklet explains the basics of using *SW* to enter, format, and edit text and mathematics. It also provides a series of step-by-step examples illustrating how to perform mathematical computations and plot mathematical expressions in *Scientific WorkPlace* and *Scientific Notebook*, and it briefly discusses how to use the Exam Builder to create algorithmically generated exams and quizzes. Finally, the booklet explains how to preview and print documents and how to use the built-in power of LaTeX with *Scientific WorkPlace* and *Scientific Word* when you need to produce a document with a finely typeset appearance.

Making Sure You Have What You Need

Before you attempt to install and use Version 3.5 of *SW*, make sure your personal computer or network client computer meets the hardware and software requirements shown below:

System Requirements	*WorkPlace* or *Word*	*Notebook*	*Viewer*
IBM-compatible PC (486 or higher)	•	•	•
RAM	16 MB	8 MB	8 MB
CD-ROM drive	•	•	
Available disk space	55 to 180 MB*	30 to 100 MB*	10-15 MB*
Windows 95, 98, 2000, or NT 4.0	•	•	•
Internet Explorer 3.01 or higher	•	•	•
Windows Multi-language Support	•	•	•

*depends on the type of hard drive and installation options

Additionally, we recommend that you have at least 60 MB of RAM in a permanent Windows swap file on your computer. See your Windows documentation if you need information about installing Windows, using Windows Setup, installing a printer, or creating a swap file. For more information about swap files, see the Technical Reference available from the General Information Index in the online Help.

Installing and Licensing SW

Important Before you open the product package, please read the license agreement that accompanies *SW*. By installing and using the product, you accept the terms of this agreement.

Scientific WorkPlace, Scientific Word, and *Scientific Notebook* are available on CD-ROM. Please remember that you may make copies of the software only for your personal use. *Scientific Viewer* is available from the MacKichan Software, Inc. website at **http://www.mackichan.com** or on the system CD.

Because the program files are compressed, you can't simply copy them to your computer. To install *SW* and its related files, you must run the installation program, which decompresses the files and copies them to your hard disk or your network. The installation program is supplied on the CD-ROM and with the downloaded program.

With Version 3.5, we are initiating a new licensing process. Once you have installed *Scientific WorkPlace, Scientific Word,* or *Scientific Notebook,* you must obtain a license to activate all the features of the program you have purchased. Until you have obtained your license, you can use only those program capabilities that are provided with the *Viewer*. You don't need a license to use the *Viewer*.

Install the Software

If you are installing the *Viewer,* you don't need a serial number for the software. Otherwise, before you begin the installation, be sure you have your serial number handy. The installation process differs for personal computers and networks. Make sure you follow the correct set of instructions.

Installing SW on Your Personal Computer

► **To install the software on a Windows computer**

1. Start Windows.
2. If you're installing *SW* from a CD-ROM, insert the CD-ROM into your CD-ROM drive. Normally, the installation program starts automatically. If it doesn't,
   a. From the Windows Start menu, choose Run.
   b. Type ***drive*:setup**, where ***drive*** is the letter of the CD-ROM drive.
   c. Choose OK.

   or

If you're installing the *Viewer* from a downloaded file:

a. From the Windows Start menu, choose Run.
b. Type ***drive*:*folder*\\SNViewer.exe**, where ***drive*** is the letter of the drive and ***folder*** is the name of the folder containing the downloaded file.
c. Choose OK.

3. Follow the instructions in the installation panels on the screen.

   If you're installing *Scientific WorkPlace* or *Scientific Word*, the system will ask you to choose between Standard LaTeX support and Multilingual LaTeX support. We recommend that you choose Multilingual LaTeX support.

4. When the program asks whether you want to restart your computer, click Yes.

Installing SW on a Network

If you are using *SW* on a network, your network administrator must first install the software on the network, according to the instructions accompanying your network license. Then, you must run the installation on your client computer.

► **To install the software on a client computer**

1. From the client computer, open the network directory containing *SW*. See your network administrator for the exact location.

2. Run `setup.exe.`

3. Follow the instructions on the screen.

4. When the program asks whether you want to restart your computer, choose Yes.

Obtain a License

To activate all capabilities and features in your *Scientific WorkPlace*, *Scientific Word*, or *Scientific Notebook* installation, you must obtain a license for the installation on your computer. Until you obtain your license, only those capabilities that are provided with the *Scientific Viewer* are available. You can obtain your license from the World Wide Web or by e-mail, fax, telephone, or letter. We recommend that you use the World Wide Web; your license will be provided to you quickly by e-mail. For client installations, see your network administrator.

To obtain a license, you must have a serial number. If you don't have a serial number, you can continue to use the *Viewer* free of charge.

► **To obtain a software license**

1. Start *SW*:
   - In the *SW* program group, double-click the *SW* icon.

     or
   - From the Windows Start menu, choose Programs, and then choose and open your *SW* product.

2. From the Help menu, choose Register.

3. Follow the instructions in the Order box to enter your name, address, and CD serial number.

4. Choose the method you want to use to register your software, and then choose OK.

5. When you receive your license file, follow the instructions provided with it to install the file.

Exploring SW

When you first start *SW*, you see a screen like this:

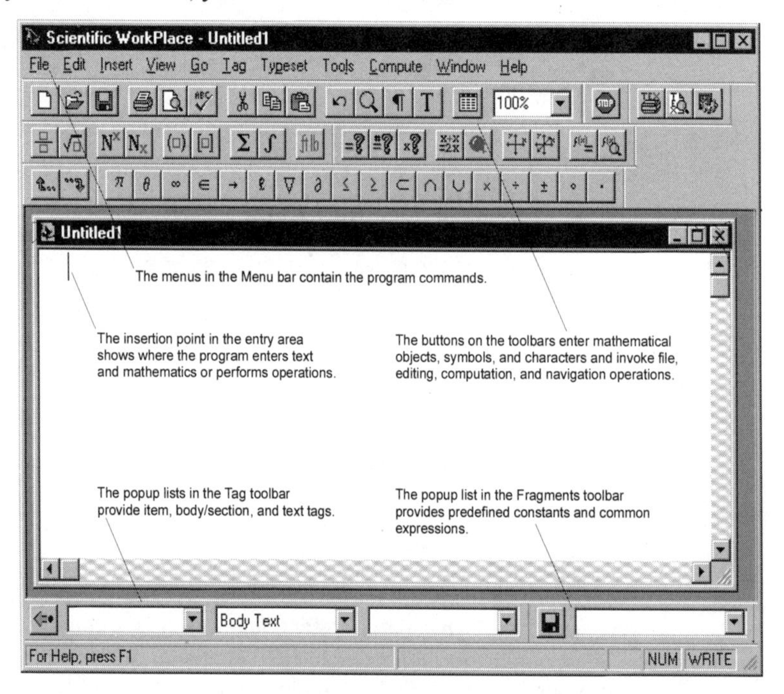

The program displays a Welcome document that contains valuable information about the software. Please read the Welcome document, then take a minute to try out the software. You'll see just how easy it is to work with text and mathematics in *SW*.

Start a New Document

Open a new document with an empty page.

1. Click [new document button].

2. From the Shell Directories list, choose General.

3. From the Shell Files list, choose Blank Document and choose OK.

Enter Text and Math

You can use natural mathematical notation to enter an expression. *SW* correctly interprets your mathematics, which appears in red on your screen.

1. Type **To integrate** and press the spacebar.
2. Click [T] to change from text to mathematics.
3. Type **x**, click [N^x], type **2**, and press the spacebar.
4. Click [M] to change back to text.
5. Press the spacebar; then type **in SW, enter** and press the spacebar.
6. Click [$\int$]. Note that *SW* changes to math automatically because it recognizes that $\int$ is mathematics.
7. Type **x**, click [N^x], type **2**, press the spacebar, and then type **dx.** On your screen, you should see this:

   To integrate x^2 in SW, enter $\int x^2 dx$

Change the Screen Appearance of Text

You can change the screen appearance of your document by changing the *tags,* or formatting instructions, applied to the content. By applying tags, you can emphasize a portion of text, such as by making it bold, italic, or large, or you can create headings, centered text, and lists.

1. Highlight SW in the expression you have entered.
2. Click the Text tag box on the Tag toolbar at the bottom of the screen.

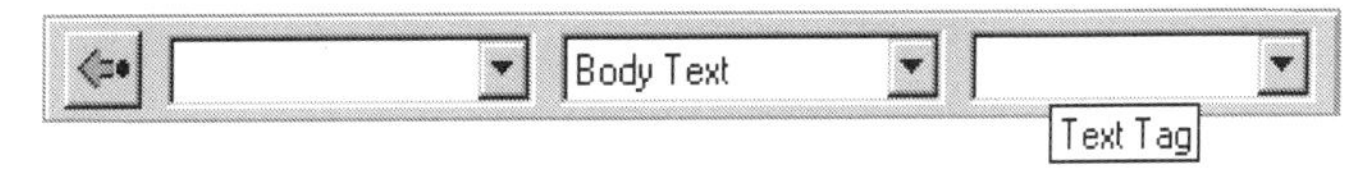

3. On the list of tags displayed, click Bold.

   The program changes SW to boldface.
4. Place the insertion point in the expression.
5. Click the Section/Body tag box at the bottom of the screen and click Body Center.

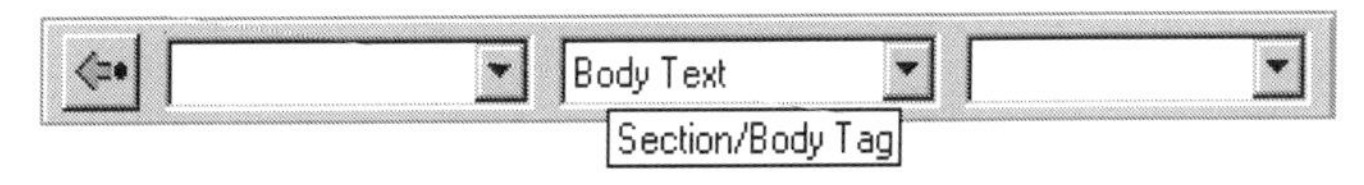

Now your expression is centered:

To integrate x^2 in **SW**, enter $\int x^2 dx$

You can assign tags to function keys for fast application or apply them from the Tag menu.

Compute

If you're using *Scientific WorkPlace* or *Scientific Notebook*, you can perform complex computations.

1. Place the insertion point at the end of the expression $\int x^2 dx$.
2. Click to evaluate the expression. The computational engine performs the integration (a special pointer shows while the computation is in progress) and places the calculated value in your document:

To integrate x^2 in **SW**, enter $\int x^2 dx = \frac{1}{3}x^3$

Plot Mathematics

With *Scientific WorkPlace* and *Scientific Notebook*, you can plot the values you compute, and then add values to those plots, all without leaving your document.

1. With the insertion point to the right of or anywhere in the expression $\int x^2 dx = \frac{1}{3}x^3$, click .

   The computational engine plots your mathematics:

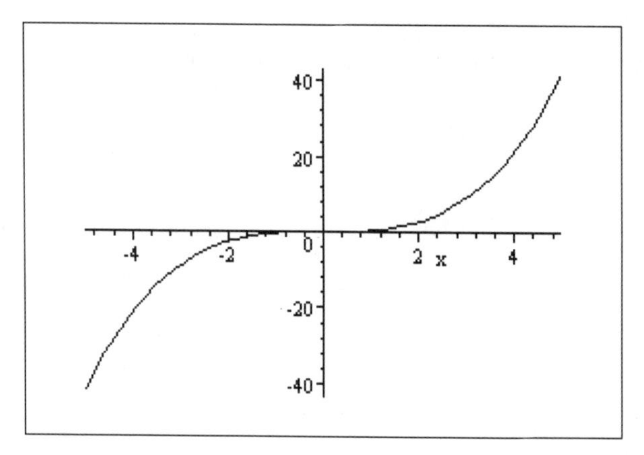

2. Now see how easy it is to add to the plot:

a. In your expression, select x^2.
b. Hold down the mouse button and drag the selection to the plot, making sure the pointer is inside the plot border. The graph is replotted, so that you now see this:

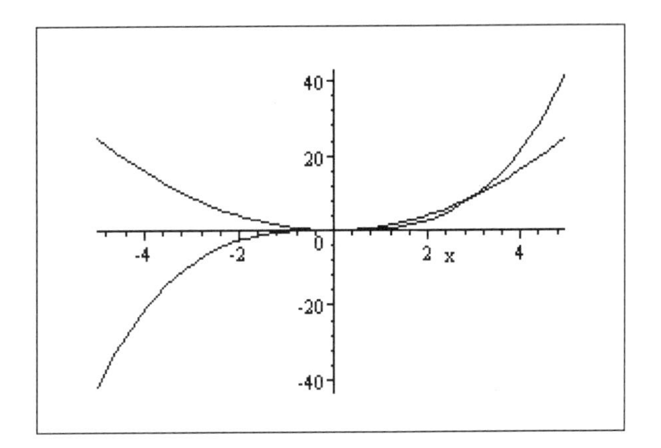

Print the Document

When you print your document with *SW* Version 3.5, your output looks like what you see on the screen. Remember, with Version 3.5 of *Scientific WorkPlace* and *Scientific Word,* you can also choose to typeset your document with LaTeX using the Typeset commands. See Chapter 4 "Typesetting Your Document" for more information.

1. Click [printer button].
2. In the Print dialog box, select the printer you want to use, and then choose OK.

Browse the Internet

If you have Internet access, you can go directly to any Internet location with a Universal Resource Locator (URL) without ever leaving your *SW* document. For example, you can visit our web site, where you can find even more information about *SW*.

1. From the File menu, choose Open Location.
2. In the Open Location dialog box, enter this URL: **http://www.mackichan.com**
3. Choose Open.

You can specify any URL on the World Wide Web. If the location you specified isn't an *SW* document, the program activates your web browser. Your *SW* document remains open. Any *SW* documents with a `.tex` extension that are available on the Internet are available as read-only documents.

Save and Close the Document

Save and close your document, unless you are using the *Viewer*.

1. Click [save icon].
2. Type a name for the document and choose OK.
3. From the File menu, choose Close.

Leave SW

You can leave *SW* several ways.

- From the File menu, choose Exit.
  or
- At the top left corner of the program window, double-click the butterfly, or click the butterfly once and then choose Close.
  or
- At the top right corner of the program window, click the Close button [X].
  or
- Press ALT+F4.
  If you haven't saved your document, the program prompts you to save your work.

2 Learning the Basics

SW is intuitive. Whether you're writing text or mathematics, you'll find that using *SW* is so easy you won't want to use pencil and paper again.

Using the Workplace

The *SW* program window is your workplace. Take a moment to familiarize yourself with the window, its menus, and its most commonly used toolbars. Remember that not all toolbars appear for all products.

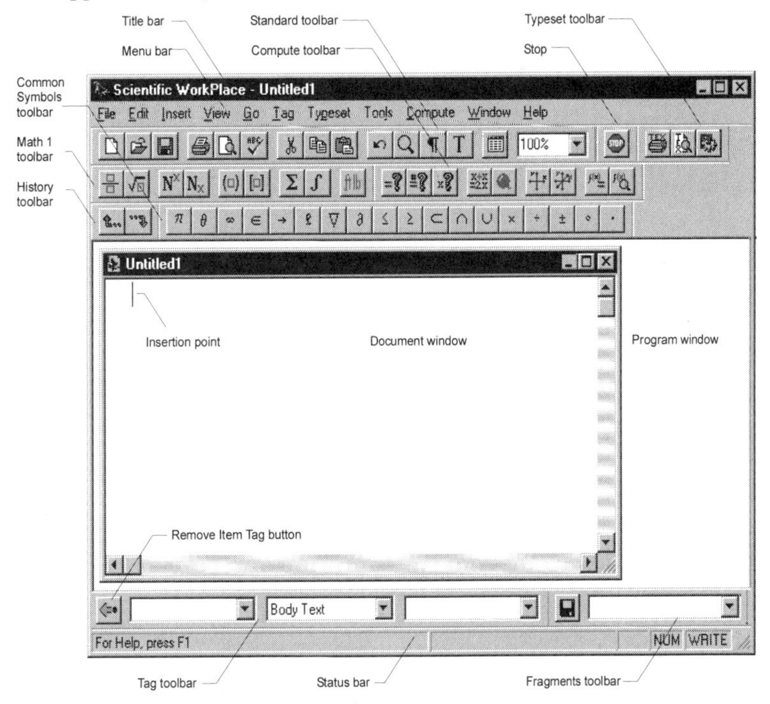

The toolbar buttons are identical in function to many of the menu commands. Point the mouse at each button briefly to display a *tooltip* identifying the button and its function. The Standard toolbar contains buttons for the most frequently used commands on the File and Edit menus:

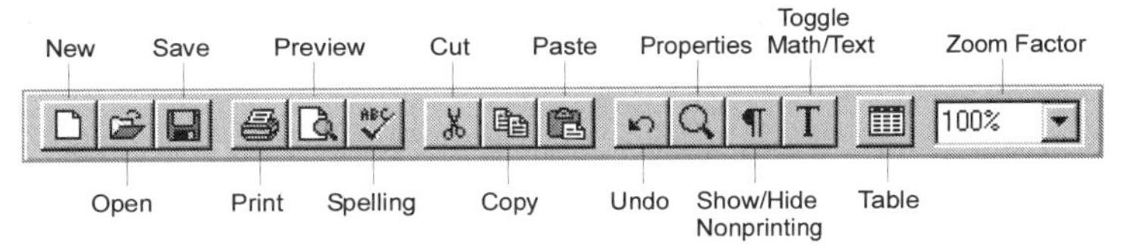

Other toolbars are available from the View menu.

► **To show or hide a toolbar**

1. From the View menu, choose Toolbars.
2. In the Toolbars dialog box, check the box next to each toolbar you want to display, and uncheck the box next to each toolbar that you want to hide.
3. Choose Close.

Opening a Document

Starting *SW* automatically opens a blank document, which is a *shell,* or template, for a typical new document. If you want to produce a similar document, you can begin entering information right away. If you want to create some other kind of document, open a new file and choose a different kind of shell. Each *SW* shell is different. Explore a little and see what's available.

► **To open a new file**

1. On the Standard toolbar, click or, from the File menu, choose New.
2. From the Shell Directories list, choose the kind of document you want.
3. From the Shell Files list, choose the shell you want for your document.
4. Choose OK.

Instead of opening a new document, you may want to open a document that already exists.

► **To open an existing file**

1. On the Standard toolbar, click or, from the File menu, choose Open.
2. Choose the file folder and name.
3. Choose OK.

Entering and Editing Text

Entering information is straightforward in *SW,* whether you enter text from the keyboard or from the special character panels. You can use standard editing tools to revise your document. Formatting the information you enter involves using tags—collections of formatting and behavior properties that determine the appearance of your document. By applying tags to your information, you can create a consistent appearance throughout your document without having to format each element individually.

Entering Text in SW

Unless you tell it otherwise, *SW* assumes that everything you type is text. When you're entering text, the Math/Text toggle on the Standard toolbar appears as

► **To enter text, just start typing.**

Most keyboards don't contain special text characters, but you can enter many of them from the panels available when you display the Symbol toolbar:

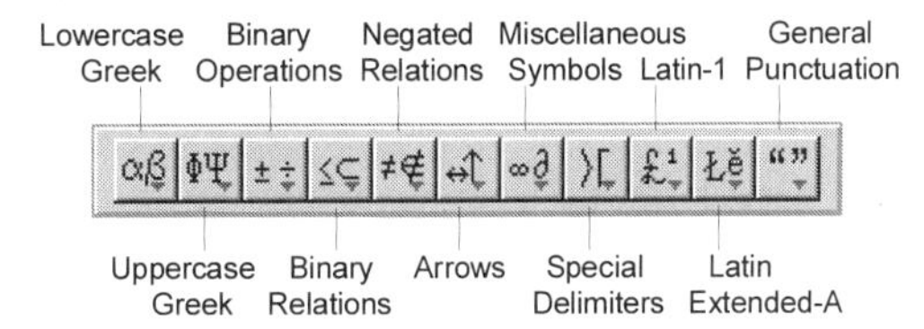

► **To enter special characters**

1. From the View menu, choose Toolbars, check Symbol, and choose Close to display the toolbar.
2. Click one the buttons on the toolbar to display the corresponding panel of special characters.

   Most special text characters appear on the Greek, Latin-1, Latin Extended-A, or General Punctuation panels. Open the panels to view the available characters.
3. Click the character you want.

► **To place information in a table**

1. On the Standard toolbar, click or, from the Insert menu, choose Table.
2. Set the number of rows and columns you want in the table. You can also change the alignment of the columns and the alignment of the table in relation to the text baseline.
3. Choose OK.
4. Fill your table with information, moving from cell to cell with the TAB key.

Editing Text

Once you've entered your text, you may need to edit it. You can use standard clipboard and drag-and-drop operations to cut, copy, paste, and delete selections. If you change your mind, you can undo the most recent change or deletion. Also, you can also use the spell check feature and the find and replace feature to help revise the content of your document. These tools are available from the Edit menu; many are also available from the Standard toolbar.

Another way to edit text in *SW* is by changing the *properties* of individual characters. For example, you might add an accent to a character so that *n* becomes *ñ*.

► To edit the properties of a character

1. Select the character or place the insertion point to its right.
2. Open the Character Properties dialog box:
    - On the Standard toolbar, click the Properties button .

      or
    - From the Edit menu, choose Properties.

      or
    - Press CTRL+F5.
3. Make the changes you want, and then choose OK.

► To undo your most recent change or deletion

- Click or, from the Edit menu, choose Undo.

Formatting with Tags

In *SW*, everything you enter carries one or more tags. *SW* automatically applies the Body Text tag to all text that you enter. By changing the Body Text tag to a different *section/body tag,* you can create a heading or a centered paragraph, and by adding an *item tag,* you can create a list. In addition to applying tags from the popup lists at the bottom of the program window, as shown here, you can apply tags with the function keys and with the Apply command on the Tag menu.

► To enter a heading

1. Click the Section/Body Tag popup list at the bottom of the program window:

2. From the tag list that pops up, click the heading level you want.
3. Type the text of the heading and then press ENTER.

▶ **To enter a list**

1. Click the Item Tag popup list at the bottom of the program window:

2. From the tag list that pops up, click the type of list you want.

3. Type the first item in the list and press ENTER. Repeat for each item in the list.

4. To complete the list, click .

With another kind of tag, called a *text tag,* you can emphasize a text selection. For example, you might want to make a selection appear Bigger, Smaller, **Bold,** *Italic,* or *Strongly Emphasized.* Depending on the text tags available in the document style, your text might appear as `Typewriter text,` Sample Text, or KEYBOARD INPUT.

▶ **To emphasize a portion of text**

1. Select the text you want to emphasize or, if you haven't typed the text, place the insertion point where you want the emphasis to begin.

2. Click the Text Tag popup list at the bottom of the program window:

3. From the tag list that pops up, click the text tag you want.

4. If you haven't yet typed the text, type it and then choose the Normal text tag to turn off the emphasis.

Changing the Screen Appearance of Tagged Text

The appearance of tagged text on the screen depends on the tag properties. The properties determine the type face, font size, font style, indentation, justification, and many other aspects of the screen appearance of tagged text. Tag properties are collected in files called *styles.* Style files have a file extension of `.cst`. The style files are also used to format your document when you print it without typesetting but they are ignored if you typeset your document.

Each document shell has an associated style file. When you choose a shell for a new document, the associated style file, with its collection of tag properties, determines the way your document appears on the screen.

You can change the screen appearance of your document by modifying the tag properties. If you want the modification to affect the screen appearance of all documents created with the style, you can save the changes to the style. Otherwise, you can leave the style (and all documents created with it) unchanged and save the changes to a new style.

► To change the screen appearance of tagged text

1. From the Tag menu, choose Appearance. The program opens the Tag Appearance dialog box.

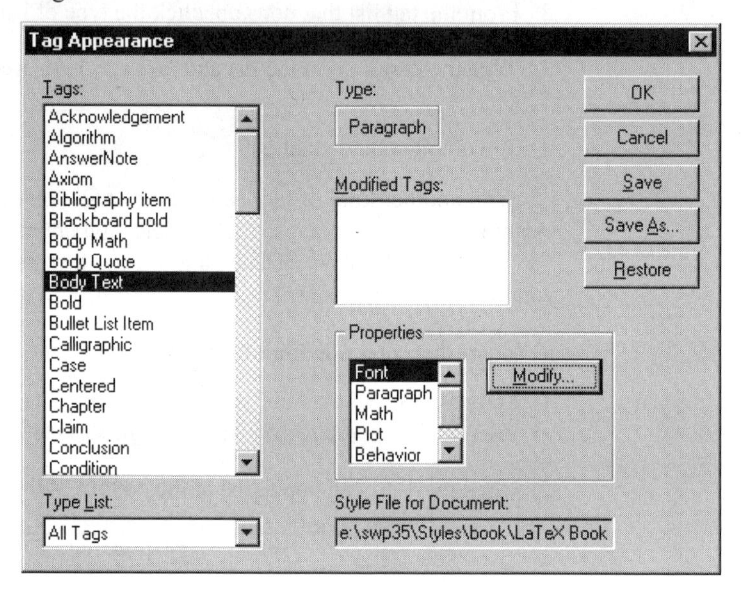

2. From the Tags list, select the tag whose screen appearance you want to change.

3. In the Properties area, double-click the property you want to change, or select the property and choose Modify.

Choose	To modify the appearance of
Font	Type face, size, style, and color
Paragraph	Justification, indentation, line spacing, background color
Lead-in	Type face, size, alignment, and spacing of lead-in objects
Math	Screen color, size, and placement of mathematical objects
Behavior	Tag for following paragraph; appearance of tag name in popup lists
Plot caption	Background color and font of caption

4. Make the changes you want in the property dialog box and then choose OK.

5. Repeat steps 2-4 for all tags whose properties you want to change.

6. Save the changes:
   - Choose Save to save the changes to the current style file, which is listed in the box labeled Style File for Document. This will alter the screen appearance of all documents created with the style.

or

- Choose Save As to save the settings in a new style file. Enter a name for the new file and choose Save.

7. Choose OK.

Formatting the Page

If you print your document without typesetting it, the margins, headers, footers, and page numbers are set initially by the document shell. These elements don't appear in the document window, but you can see them if you preview your document. You can change them to suit your needs.

▶ **To modify the page setup specifications**

1. From the File menu, choose Page Setup.
2. Choose the tab you need:

Choose	To
Margins	Set the left, right, top, and bottom page margins
Headers/Footers	Specify headers, footers, and page numbering
Counters	Specify the page numbering style

3. Choose OK.

If you typeset your document in *Scientific WorkPlace* or *Scientific Word,* the Page Setup specifications are ignored, and the margins, headers, footers, and page numbers are set according to the LaTeX typesetting specifications.

Entering and Editing Mathematics

Because *SW* assumes you're entering text, you must tell it when you want to enter mathematics. Then, you can enter mathematics easily using the toolbar buttons, Insert menu commands, or keyboard shortcuts.

▶ **To start mathematics**

- Click [T] or, from the Insert menu, choose Math.

When mathematics is active, the Math/Text toggle appears as [M].

▶ **To return to text**

- Click [M] or, from the Insert menu, choose Text.

When text is active, the Math/Text toggle appears as [T].

Entering Mathematical Characters

▶ To enter a mathematical character

- Click the character you want from the Common Symbols toolbar:

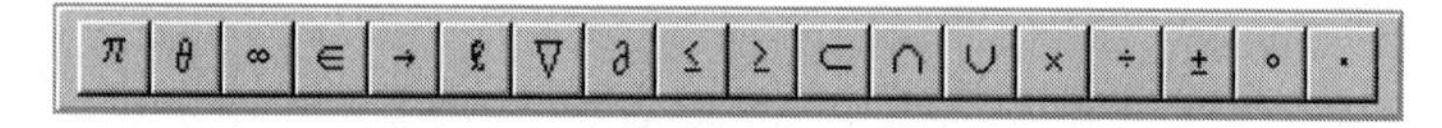

or

- Click one of the buttons on the Symbol toolbar to open a panel of special characters, and then click the character you want.

We suggest you open the symbol panels one by one to explore all the available characters. Choosing a mathematical character automatically starts mathematics, even if you have not toggled to mathematics.

Entering Mathematical Objects

Mathematical objects such as fractions, radicals, subscripts and superscripts, operators, and brackets are available from the Insert menu and the Math toolbars:

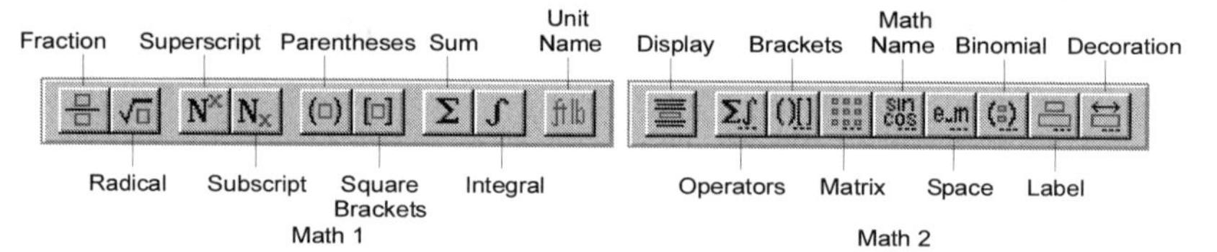

Also, you can enter many mathematical objects using keyboard shortcuts (see Appendix B).

When you enter a mathematical object, the program automatically starts mathematics and places a *template* for the object in your document. You complete the object by filling in the template.

Here are instructions for entering some common mathematical objects: fractions, subscripts and superscripts, expressions in parentheses, matrices, and operators such as $\int$ and $\sum$. Here too are instructions for creating in-line and displayed mathematics.

▶ To enter a fraction

1. Click or, from the Insert menu, choose Fraction.

   On the screen, you see $\frac{\square}{\square}$, and the Math/Text toggle changes to **M**.

2. Type the numerator, and then press TAB.

3. Type the denominator, and then press the spacebar.

▶ To enter a superscript or subscript

1. Click [T] or, from the Insert menu, choose Math to start mathematics.
2. Type a variable.
3. Click [N^x] to enter a superscript or [N_x] to enter a subscript.
4. Type the superscript or subscript, and then press the spacebar.

▶ To enter an expression in parentheses

1. Click [(□)].
2. Type the expression. Notice that the parentheses are elastic—they expand horizontally and vertically as far as necessary to enclose the expression you enter.
3. Press the spacebar.

▶ To enter a matrix

1. Click [matrix] or, from the Insert menu, choose Matrix.
2. Set the number of rows and columns you want.
3. Choose OK.
4. Fill your matrix with mathematics, moving from cell to cell with the TAB key. You can change the alignment of the matrix, just as you can for a table.

▶ To enter an operator

1. Click [Σ∫] or, from the Insert menu, choose Operator.
2. Double-click the operator you want.
3. Click [N_x] and then type the lower limit for the operator.
4. Press TAB, and then type the upper limit.
5. Press the spacebar, and then type the variable.
6. If the variable carries a subscript, click [N_x], type the subscript, and then press the spacebar.

If the expression is in a line of text, the limits are automatically placed to the right, like this: $\sum_{i-1}^{n} a_i$. If the expression is displayed on a line by itself, the limits are automatically placed above and below the operator:

$$\sum_{i-1}^{n} a_i$$

► To change in-line mathematics to displayed mathematics

1. Enter a mathematical expression, and then select it.
2. Click .

► To change displayed mathematics to in-line mathematics

- Place the insertion point to the right of the display, and then press BACKSPACE.

Entering Mathematics with Fragments

If you enter a certain expression or equation frequently, you can save it as a *fragment* and then enter it in any document with just a few keystrokes. Fragments are available from the File menu and from the popup list on the Fragments toolbar at the bottom of the program window:

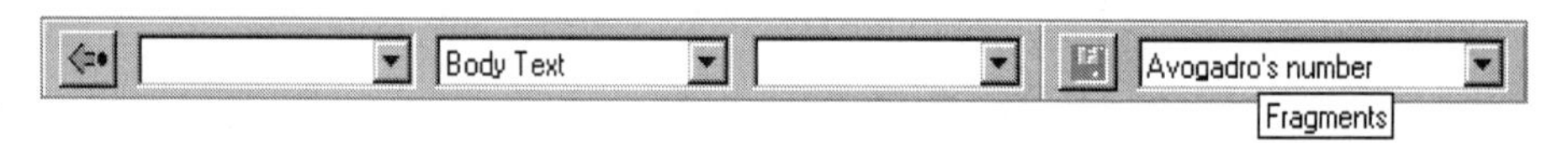

You can save both mathematics and text in a fragment. The only restriction is that a fragment can be no larger than a paragraph.

► To enter a fragment in your document

- Click the Fragments popup to display the list of available fragments, and then click the name of the fragment you want.
  or
- From the File menu, choose Import Fragments, select the fragment you want, and then choose OK.

► To save an expression as a fragment

1. Select the expression or equation you want to save as a fragment.
2. On the Fragments toolbar, click or, from the File menu, choose Save Fragment.
3. Type a file name to be used to recall the fragment. Avoid using the name of a TEX command for the fragment.

4. Leave the directory unchanged.

5. Choose OK.

SW saves your fragment in the `Frags` directory of your program installation and inserts its name in the list of available fragments.

Using Body Math

You can enter a series of mathematical expressions even faster if you enter them in a Body Math paragraph. The Body Math tag is available from the Section/Body Tag popup list. Each time you press ENTER in a Body Math paragraph, the program automatically switches to mathematics. This makes it easy to use your document as a mathematics scratchpad.

Editing Mathematics

When you need to edit your mathematics, you can use standard clipboard and drag-and-drop operations to cut, copy, paste, and delete. You can also edit the properties of mathematical characters, symbols, and objects.

When you edit the properties of a mathematical character or symbol, *SW* opens the Character Properties dialog box. When you edit the properties of a mathematical object, *SW* opens a context-sensitive dialog box; that is, a dialog box that corresponds to the mathematical object you've selected. If you haven't selected an object, the program opens the dialog box appropriate for the object to the left of the insertion point.

► To edit the properties of a mathematical object

1. Select the object or place the insertion point to its right.

2. Open a context-sensitive Properties dialog box:

   - On the Standard toolbar, click the Properties button .
     or
   - From the Edit menu, choose Properties.
     or
   - Press CTRL+F5.

3. Make the changes you want, and then choose OK.

Saving Your Document

When you've finished writing, save your *SW* document.

► To save your document

1. Click or, from the File menu, choose Save.

2. Type a name for your document and choose OK.

   Any file name compatible with your Windows system is acceptable.

SW saves your document as a .tex file. However, if you're working in *Scientific WorkPlace* or *Scientific Word*, you can output your document using a new output filter called Portable LaTeX. This new filter creates documents that have a .tex extension but are more easily read by standard latex installations. When you save a document as a Portable LaTeX file, the program doesn't insert the line **\input{tcilatex}** in your document, nor does it include any LaTeX packages that aren't part of a standard LaTeX installation. Portable LaTeX is unavailable for Style Editor styles and for styles created under LaTeX 2.09.

► **To save your document as a Portable LaTeX file**

1. From the File menu, choose Save As.
2. Type the name of the file.
3. In the Save as type area, select Portable LaTeX (*.tex).
4. Choose Save.

Previewing and Printing Your Document

You can print documents from the document window or the preview window. For information about typeset previewing and typeset printing, see Chapter 4 "Typesetting Your Document."

► **To preview a document**

1. Click or, from the File menu, choose Preview.
2. Use the scroll bars and the toolbar buttons to examine your document in the preview window.
3. When you're ready to leave the previewer, click the Close button.

► **To print a document**

1. From the document window, click or, from the File menu, choose Print.

   or

   From the Menu bar of the preview window, choose Print.
2. Specify the printer and the print options you want and choose OK.

SW uses the same routines to preview and print your document as it does to display it in the document window. Therefore, what you see when you preview or print your document without typesetting it is similar to what you see when you display it, except that margins, headers, footers, and page numbers aren't displayed in the document window.

Browsing with SW

With *SW*, you can access information located elsewhere in the same document or in other documents on the current system or network. If you have an Internet Service Provider, you can access information on the Internet without ever leaving your *SW* document. You can open any URL on the World Wide Web or jump to any Internet location that is defined in a *hypertext link,* or jump, to information in another location.

Jumping with Hypertext Links

Hypertext links in *SW* documents appear in green by default in the program window. Any time you encounter a link in an *SW* document, you can jump immediately to the linked information, whether that information is in the same document, in other documents on the current system or network, or on the Internet.

► **To jump to the target of a hypertext link**

1. Place the insertion point in the link so that the hypertext pointer appears.
2. From the Tools menu, choose Action.

   or

   If the link appears in a document saved as read-only (such as those provided with the program), click the link.

   or

   If the link appears in a document not saved as read-only, press CTRL while you click the link.

The program moves the insertion point to the specified location, opening specified documents or linking to the Internet as necessary.

Browsing the Internet

In addition to jumping to any Internet location defined in a hypertext link, you can open any URL on the World Wide Web.

► **To open an Internet location**

1. From the File menu, choose Open Location.
2. In the Open Location dialog box, enter the URL of the location you want to access.
3. Choose Open.

If the location you specify isn't an *SW* document, the program activates your web browser. *SW* documents with a `.tex` extension that are placed on the Internet are available as read-only documents. You must save a copy of them locally if you want to use or perform computations on any information they may contain. Your *SW* document remains open while you browse.

► **To cancel an attempt to open an Internet location**

- On the Stop toolbar, click the Stop Operation button  or press CTRL+BREAK.

Creating Hypertext Links

Hypertext links have two parts: the link and the target. You must create both parts to complete the link. What appears in your document at the point of the link can be text or a graphic. (Note that because hypertext is intended for use online rather than in print, special attention is required to typeset documents containing hypertext links. More information appears in the online Help.)

The target of a hypertext jump can be any object to which you've assigned an identifying key or marker—such as a figure, a section, or equation—or any file, including files on the Internet. The address of the target differs depending on the target itself and follows the model used in standard web browsers.

► **To create a hypertext link**

1. Place the insertion point where you want the link to appear.
2. From the Field toolbar, click or, from the Insert menu, choose Field, and then choose Hypertext Link.
3. In the Hypertext Link dialog box, specify the text or graphic that will signal the link in the document window or when you print without typesetting. See the online Help for information about typesetting documents containing hypertext links.
4. In the Targets box, enter the address of the target.
5. Choose OK.

When you create links, make sure the target exists, whether it is a document or a marker within a document on the current system or on the Internet.

► **To create a marker in an SW document**

1. Open the target document and place the insertion point where you want the marker.
2. On the Field toolbar, click the Marker button or, from the Insert menu, choose Field, and then choose Marker.
3. In the Key box, enter a unique key for the item.
4. Choose OK.
5. Save the document.

Managing Your Documents

SW documents are associated with many files in addition to the one that contains the document itself. Some of these files may contain graphics, subdocuments, or style information. Others are created by the program when you typeset with LaTeX. Depending on the file type, *SW* may not store an associated file in the same directory that holds the document file. Therefore, when you copy, delete, or rename a document and especially when you exchange documents with colleagues, use the *SW* Document Manager to ensure that all the files associated with your document are handled correctly.

With the Document Manager, you can copy, delete, rename, view, or clean up a document. Also, you can *wrap* or *unwrap* a document. That is, you can gather together into a single text file all those files that accompany a document, or you can break a file that has been wrapped into separate files again. Wrapping a file before you send it to another location or by e-mail or on diskette ensures that all necessary files are sent along with the primary document file. You can unwrap a file with *SW*, the Document Manager, or an ASCII editor.

▶ To start an SW Document manager operation

1. From the File menu, choose Document Manager.
2. Choose a document.
3. Choose the operation you want.

▶ To wrap a document

1. From the File menu, choose Save As.
2. In the box labeled Save as type, choose Wrap (*.rap).
3. Choose Save.

or

1. From the File menu, choose Document Manager.
2. In the File Selection box, type the name of the document you want to wrap.
3. Choose Wrap.
4. If you need to wrap the document for a different system:
   a. Choose Browse.
   b. In the Save as type area, choose the type of wrap file appropriate for the system on which the document will be unwrapped.
   c. Choose Save.
5. Exclude any files you don't want to wrap with your document.

   If you are sending the document to someone who has *SW*, remember that you don't need to send plot snapshots or typesetting specifications.

6. Choose OK.
7. The Document Manager creates a file with the same name as your document and the extension you selected. The file contains your document, the additional files you included, and instructions for using an ASCII editor to recreate the original files, in case the recipient uses different software.
8. When the operation is complete, choose OK.
9. Choose Close.

► To open a wrapped document in SW

1. From the File menu, choose Open.
2. Select the name and location of the wrapped file.
3. Set the wrapped file type
4. Choose OK.

► To unwrap a document with the Document Manager

1. From the File menu, choose Document Manager.
2. In the File Selection box, type the name of the document you want to unwrap.
3. Choose Unwrap.
4. Exclude any files you don't want to unwrap with the document.
5. Choose OK.
6. The Document Manager unwraps the document, placing each file in the correct directory.
7. When the operation is complete, choose OK.
8. Choose Close.

► To unwrap a document with an ASCII editor

1. Open the wrapped file with the editor.
2. Follow the instructions in the file header.

Customizing the Program

SW is flexible: you can customize it to suit the way you work. By modifying the appearance of the program and document windows and the use of program tools and defaults, you can make *SW* even more convenient to use.

Changing the Appearance of the Windows

In addition to sizing the program window to your liking, you can modify it by displaying only those toolbars and symbol panels you use most frequently and by moving toolbars to screen locations that are convenient for you.

▶ To display or hide toolbars

1. From the View menu, choose Toolbars.
2. Check the toolbars you want to display and uncheck the toolbars you want to hide.
3. Choose Close.

▶ To return to the original toolbar display

1. From the View menu, choose Toolbars.
2. Choose Reset and then choose Close.

▶ To move a toolbar to a new location

1. Display the toolbar.
2. Place the mouse pointer anywhere in the gray area surrounding the toolbar buttons.
3. Drag the toolbar to a new location.

   You can *dock* the toolbar at the top, bottom, or sides of the program window, or you can let it *float* on your desktop or in the entry area of the program window.

▶ To float a symbol panel on the screen

1. On the Symbol toolbar, click the symbol panel you want.
2. Place the mouse pointer on the title bar of the panel.
3. Drag the panel to a new location on the screen.

▶ To reshape a toolbar

1. Float the toolbar on the screen.
2. Drag any side of the toolbar to reshape it.

For faster access to symbols and characters, you can leave open the symbol panels you use most often and float them anywhere you want.

► To close a symbol panel

- In the upper-right corner of the symbol panel, click the Close button ☒.
  or
- On the Symbol toolbar, click the button for the symbol panel.

In *SW*, you can have several document windows open at the same time, and you can arrange them conveniently within the program window. You can set the magnification and the characteristics of the view separately for each window.

► To open a document in a new window

- Open an existing document:

  a. On the Standard toolbar, click or, from the File menu, choose Open.
  b. Specify the file you want to open, and then choose OK.

  or

- Open a new document

  a. On the Standard toolbar, click or, from the File menu, choose New.
  b. From the New dialog box, choose the shell you want, and then choose OK.

  or

- Open another view of the active document
  - From the Window menu, choose New Window.

► To arrange the open document windows

- From the Window menu, choose Cascade, Tile Horizontally, or Tile Vertically.
  or
- With the mouse, drag the Title bar of a document window to the position you want.

You can change the size of the document in each window from 50% to 400% of normal size.

► To change the magnification in the active window

- From the View menu:
  - Choose 100% or 200%.
    or
  - Choose Custom, set the percentage of magnification you want, and choose OK.

  or

- On the Standard toolbar, click the Zoom Factor box 100%, and then choose the percentage of magnification you want or type it and press ENTER.

Changing the Tools and Defaults

Working in *SW* is fast and convenient, but you can make it even more so if you set the function keys to apply the tags you use most often and set the program defaults to customize the way the program works.

Initially, the function keys have these tag assignments:

Key	Tag	Key	Tag
F2	Remove Item Tag	F7	Numbered List Item
F3	Body Text	F8	Bullet List Item
F4	Normal	F9	Calligraphic
F5	Bold	F11	Section
F6	Emphasize	F12	Subsection

You can set global function key assignments that apply to all documents, or you can override the global settings with different function key assignments for each style.

▶ To change a function key assignment

1. From the Tag menu, choose Function Keys.
2. In the Tag Key Assignments dialog box, select the tag you want to assign to a function key.
3. In the Save with box, select the environment for the setting as global or style.
4. Position the insertion point in the box marked Press New Keys, and then press the function key you want to use for the tag.

   You can use modifiers such as CTRL, ALT, and SHIFT.
5. Choose Assign.

   If the function key you choose is already assigned to a tag, the program clears the old assignment.
6. Choose Close.

▶ To clear a tag assignment

1. In the Tag Key Assignments dialog box, select the tag whose assignment you want to clear.
2. In the Current Assignments box, select the assignment, and then choose Remove.
3. Choose Close.

By changing the User Setup defaults, you can customize the way *SW* works with files, text, mathematics, and graphics.

► **To customize a program default**

1. From the Tools menu, choose User Setup.

2. Choose the tab for the kind of default you want to set:

These defaults	Relate to
General	how the program operates internally
Edit	how certain keys and the mouse function as you enter information
Start-up Document	which document shell is displayed when you open *SW*
Graphics	how *SW* treats new graphics
Files	where files are located; how and when you save files
Math	how the Math/Text toggle and other mathematics controls operate
Font Mapping	which fonts are used to display characters not in the standard ASCII range

3. Specify the setting you want by checking or unchecking boxes and buttons, entering numbers to indicate settings, or typing information in the dialog boxes.

4. Choose OK.

3 Computing and Plotting

With just a few mouse clicks, you can perform basic and complex mathematical computations right in your document. In *Scientific WorkPlace* and *Scientific Notebook*, you can use the computational engine to perform symbolic computations fundamental to algebra, trigonometry, and calculus—evaluating, factoring, combining, expanding, and simplifying terms and expressions containing integers, fractions, and real and complex numbers. You can also perform integration, differentiation, matrix and vector operations, standard deviations, and many other more complex computations involved in calculus, linear algebra, differential equations, and statistics.

You can manipulate the results of your computations, using them to perform additional computations or plotting the results. You can plot additional items by dragging them onto an existing plot. You can build a series of expressions that show a step-by-step approach to a problem solution by computing in place—performing computations within an expression, rather than on an entire expression. And you can perform computations on the mathematics you enter or on data files you import from your calculator. Finally, you can combine the power of the computational engine with the power of the Exam Builder to create algorithmically-generated, computer-graded course materials.

Use the Settings command on the Compute menu to toggle between a full and a partial menu of computational commands. You can find extensive information about performing computations and generating plots in the online Help and in the manual *Doing Mathematics with Scientific WorkPlace and Scientific Notebook.*

▶ To perform a computation or plot a graph

1. Enter a mathematical expression.
2. With the insertion point in or at the immediate right of the expression, choose the command you want from the Compute menu or the Compute toolbar:

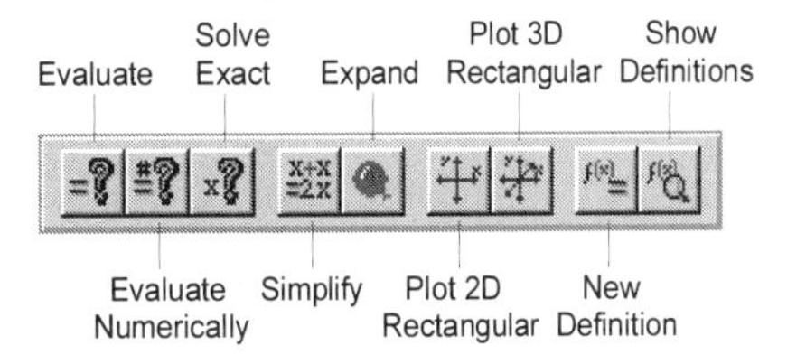

While the computation takes place, the program displays a computational pointer. Then, the program inserts the answer in your document. Most computations are fast, but some take several minutes. Occasionally, you may want to cancel a computation.

▶ To stop a computation

- Click [STOP] or press CTRL+BREAK.

Evaluate and Evaluate Numerically

You can easily collect terms, find the sum or difference of polynomials, change a quotient to rational form, or raise a number to a power with the Evaluate and Evaluate Numerically commands. When you evaluate an expression, the computational engine returns an exact or symbolic answer whenever appropriate and a numerical approximation otherwise. You can use the Evaluate Numerically command to force a numerical result for any evaluation.

Evaluate $\sqrt{2.36}$

1. Click . The program starts mathematics automatically.
2. Type **2.36**. Your screen shows the expression $\sqrt{2.36}$.
3. Click . The computational engine evaluates and you see this result:

$$\sqrt{2.36.}=1.5362$$

Evaluate $(0.16)^{-1}$

1. Click , type **0.16**, and press the spacebar.
2. Click and type **-1**. Your screen shows the expression $(0.16)^{-1}$.
3. Click . The computational engine evaluates the expression:

$$(0.16)^{-1}=6.25$$

Evaluate Numerically $936 \div 14$

1. Click or, from the Insert menu, choose Math to start mathematics.
2. Type **936**, click , and type **14**.
3. Click . Your screen displays this result:

$$936 \div 14=66.857$$

Evaluate π Numerically

1. Click [π button] and click [#? button]. The evaluation yields this result: $\pi = 3.1416$.
2. Now see how the results differ when you change the numerical accuracy with which the computational engine operates:
   a. From the **Tools** menu, choose **Engine Setup**.
   b. Choose the **General** tab.
   c. Change the setting for **Digits Used in Computations** to 50.
   d. Choose **OK**.
   e. From the **Tools** menu, choose **Computation Setup**.
   f. Choose the **General** tab.
   g. Change the setting for **Digits Shown in Results** to 50.
   h. Choose **OK**.
   i. Press ENTER, click [π button], and then click [#? button]. Your screen now shows this value for π:

$$\pi = 3.141\,592\,653\,589\,793\,238\,462\,643\,383\,279\,502\,884\,197\,169\,399\,375\,1$$

3. Return to the original settings unless you routinely need this degree of accuracy.

Evaluate these expressions

Compute these expressions with **Evaluate**:

$$27 + 33 - 16 \qquad 14.2 \times 83.5 \qquad |-11.3| \qquad (3x^2 + 3x) + (8x^2 + 7)$$

Compute these expressions with **Evaluate Numerically**:

$$\frac{8}{9} \qquad \sqrt{2} \qquad \int_0^1 e^{x^2}\,dx$$

Compute these expressions with **Evaluate** and **Evaluate Numerically**. Note the different results.

$$\frac{-\frac{2}{3}}{\frac{8}{7}} \qquad \int x dx = \frac{1}{2}x^2 \qquad \frac{5}{8} \times \frac{1}{7} \qquad (x+3) + (x-y) \qquad 4^{-3}$$

Factor

You can use the Factor command when you need to find the elements or quantities whose multiplication gives a certain product or polynomial. You can also use Factor to rationalize a denominator.

Factor 12345

1. Click T or, from the Insert menu, choose Math to start mathematics.
2. Type **12345.**
3. From the Compute menu, choose Factor. Your screen shows this result:

$$12345=3\times5\times823$$

Factor $x^2 - y^2$

1. Click T or, from the Insert menu, choose Math to start mathematics.
2. Type **x**, click N^x, type **2**, and press the spacebar.
3. Type **-y**, click N^x, and type **2**.
4. From the Compute menu, choose Factor. Your screen shows this result:

$$x^2-y^2=-(y-x)(y+x).$$

Factor these expressions

$24!$ $\quad x^6 - y^6 \quad \frac{1}{2}x^2 + 3x - \frac{20}{9}$

$5x^5 + 5x^4 - 10x^3 - 10x^2 + 5x + 5$

Combine

In addition to combining the items in mathematical expressions, you can use the Combine command to demonstrate the laws of exponents and the properties of logarithms.

Combine $\sin x \cos y + \cos x \sin y$

You don't need to type any spaces as you enter the functions. The program interprets the trigonometric functions and formats the expression correctly, displaying the function names in gray.

1. Click [T] or, from the Insert menu, choose Math to start mathematics.
2. Type **sinxcosy**.
3. Type **+cosxsiny**.
4. From the Compute menu, choose Combine, and then choose Trig Functions. This is the result:

$$\boxed{\sin x \cos y + \cos x \sin y = \sin(y + x)}$$

Combine the powers of $10^x 10^y$

1. Click [T] or, from the Insert menu, choose Math to start mathematics.
2. Type **10**, click [N^x], type **x**, and then press the spacebar.
3. Type **10**, click [N^x], and type **y**.
4. From the Compute menu, choose Combine, and then choose Powers. After the calculation is complete, your screen shows this result:

$$\boxed{10^s 10^y = 10^{s+y}}$$

Combine $\log x + \log y$

Note that the program automatically interprets *log* as a function, displaying it in a different color on your screen.

1. Click [T] or, from the Insert menu, choose Math to start mathematics.
2. Type **logx+logy**.

3. From the Compute menu, choose Combine.
4. Choose Logs. This is the result:

$$\log x+\log y=\ln xy$$

Combine the terms in these expressions

$(e^x)^y$ $3\ln x$ $\sin x \sin y$ $(2^x)^y$

Expand

You can expand the products or powers of polynomials with the Expand command.

Expand $\frac{18229}{94}$

1. Click .
2. Type **18229**, press TAB, and type **94**.
3. Click to obtain this result:

$$\frac{18229}{94}=193\frac{87}{94}$$

Expand $\tan(a+b)$

1. Click **T** or, from the Insert menu, choose Math to start mathematics.
2. Type **tan**. The program automatically recognizes the trigonometric function and displays it on your screen in gray.
3. Click (□) and type **a+b**.
4. Click . After the computation, you see this result:

$$\tan(a+b)=\frac{\tan a+\tan b}{1-\tan a\tan b}$$

Expand these expressions

$(3x^2+3x)^3$ $\sin(x+y)$ $(3x^2+3x)(8x^2+7)$ $(x+y)^4$

Simplify

You can reduce expressions to standard form, collecting like terms in addition and canceling common factors in division. When possible, the computational engine performs indicated operations exactly or reduces them to lower-level operations.

Simplify $4a+7b-(2a+b)$

1. Click or, from the Insert menu, choose Math to start mathematics.
2. Type **4a+7b-**, click , type **2a+b**, and click . The computational engine returns this answer:

$$4a+7b-(2a+b)=2a+6b$$

Simplify $\left(x^2-6x+\left(\frac{-6}{2}\right)^2\right)+\left(y^2+10y\left(\frac{10}{2}\right)^2\right)$

1. Enter the first expression:
   a. Click , type **x**, click , type **2**, and press the spacebar.
   b. Type **-6x+**, click , and then click .
   c. Type -6, press TAB, type **2**, and press the spacebar twice.
   d. Click , type **2**, and press the spacebar twice.
2. Type the second expression:
   a. Type + , click , type **y**, click , type **2**, and press the spacebar.
   b. Type **+10y**, click , and then click .
   c. Type **10**, press TAB, type **2**, press the spacebar twice, click , and type **2**.
3. Click . The computational engine simplifies the mathematics:

$$\left(x^2-6x+\left(\frac{-6}{2}\right)^2\right)+\left(y^2+10y\left(\frac{10}{2}\right)^2\right)=x^2-6x+9+y^2+250y$$

Simplify these expressions

$\sqrt[3]{8}+3$ $\qquad \sin^2 x+\cos^2 x$ $\qquad \int_1^a \frac{1}{t}dt$ $\qquad 14\frac{5}{9}$ $\qquad \frac{3x^2+3x}{8x^2+7}+\frac{5x^2+3}{2x^2+x+7}$

Check Equality

You can determine whether an equality is true or false or whether, if the test is inconclusive, the inequality may be true or false.

Check the equality $1+1=3$

1. Click [T] or, from the Insert menu, choose Math to start mathematics.
2. Type **1+1=3**.
3. From the Compute menu, choose Check Equality. You see this result:

$1+1=3$ is false

Check the equality $\frac{9}{8}-\frac{8}{9}=\left|\frac{9}{8}-\frac{8}{9}\right|$

1. Click [fraction button], type **9**, press TAB, type **8**, and press the spacebar.
2. Type - , click [fraction button], type **8**, press TAB, type **9**, and press the spacebar.
3. Type = , click [brackets button], choose [absolute value], and choose OK.
4. Type the two fractions again.
5. From the Compute menu, choose Check Equality. This is the result:

$\frac{9}{8}-\frac{8}{9}=\left|\frac{9}{8}-\frac{8}{9}\right|$ is true

Check the equality $\arcsin(\sin x)=x$

1. Click [T] or, from the Insert menu, choose Math to start mathematics.
2. Type the function **arcsin.** The program displays the function in gray on your screen.
3. Click [parentheses button], type **sinx**, press the spacebar, and type **=x**.
4. From the Compute menu, choose Check Equality. Inadequate information is provided for the computational engine to determine whether the equality is true or false. You see this result:

$\arcsin(\sin x)=x$ is undecidable

Check these equalities

$$e^{i\pi} = -1 \qquad \pi = 3.14 \qquad \sin^2 x + \cos^2 x = 1$$

Solve Exact and Numeric

As with evaluation, when you solve an expression, the computational engine returns an exact or symbolic answer whenever appropriate. You can use the Solve Numeric command to force a numerical result for any solution.

You can solve equations with one or more variables. You can solve inequalities with Solve Exact. And you can solve a system of equations if you place the equations in a matrix cell and then apply the Solve Exact command to the matrix.

Solve Exact $3x + 5 \leq 5x - 3$

1. Click T or, from the Insert menu, choose Math to start mathematics.
2. Type **3x+5**.
3. On the Common Symbols toolbar, click ≤.
4. Type **5x-3**.
5. Click x? or, from the Compute menu, choose Solve, and then choose Exact. On your screen, you see this solution:

$3x + 5 \leq 5x - 3$, Solution is: $\{4 \leq x\}$

Solve Exact $2x + 3y = 1$

1. Click T or, from the Insert menu, choose Math to start mathematics.
2. Type **2x+3y=1**. Remember that you don't need to enter any spaces; the program spaces the expression correctly.
3. From the Compute menu, choose Solve, and then choose Exact.
4. In the Solution Variables box, enter **x** and choose OK. The computational engine presents this solution:

$2x + 3y = 1$, Solution is: $\{x = -\frac{3}{2}y + \frac{1}{2}\}$

Solve Numeric $x^2 - 2 = 0$

1. Click [T] or, from the Insert menu, choose Math to start mathematics.
2. Type **x**, click [N^x], type **2**, and press the spacebar.
3. Type **-2=0**.
4. From the Compute menu, choose Solve, and then choose Numeric. The computational engine returns this numeric solution:

$$x^2 - 2 = 0, \text{ Solution is:}\{x = -1.4142\},\{x = 1.4142\}$$

Solve a system of equations

1. Click [matrix] or, from the Insert menu, choose Matrix.
2. Specify 2 rows and 1 column, and then choose OK.
3. Type **2x+3y=41** and press TAB.
4. Type **x+y=146** and press the spacebar. You see this matrix:

$$\begin{matrix} 2x + 3y = 41 \\ x + y = 146 \end{matrix}$$

5. From the Compute menu, choose Solve, and then choose Exact. The computational engine solves the system of equations in the matrix and displays this result:

$$\begin{matrix} 2x + 3y = 41 \\ x + y = 146 \end{matrix}, \text{ Solution is:}\{y = -251, x = 397\}$$

Solve these expressions

Use Solve Exact to solve these relations:

$\frac{1}{x} + \frac{1}{y} = 1$ (for x) $\quad x^2 - 5x + 4 = 0 \quad \frac{7 - 2x}{x - 2} \geq 0 \quad \begin{bmatrix} 2x + y = 5 \\ 3x - 7y = 2 \end{bmatrix}$

Use Solve Numeric to solve these relations:

$16 - 7y = 10y - 4 \quad x^5 - 5x^4 + 3x + 4 = 0 \quad \begin{bmatrix} \sin x = \cos x \\ x \in (9, 12) \end{bmatrix}$

Work with Polynomials

Use the commands on the Polynomials and Calculus submenus to collect and order the terms in polynomials, divide polynomials, determine roots, and write a rational expression as the sum of partial fractions.

Collect the terms in $3x - 7x^2 + 8x - 3 + x^5$

1. Click [T] or, from the Insert menu, choose Math to start mathematics.
2. Type **3x-7x**, click [Nˣ], type **2**, and press the spacebar.
3. Type **+8x-3+x**.
4. Click [Nˣ], type **5**, and press the spacebar.
5. From the Compute menu, choose Polynomials, and then choose Collect.
6. Type **x** in the Need Polynomial Variable dialog box and choose OK. The computational engine collects the terms and displays this result:

$$3x - 7x^2 + 8x - 3 + x^5 = 11x - 7x^2 - 3 + x^5$$

Order the terms in $3x - 7x^2 + 8x - 3 + x^5$

1. Follow the steps in the preceding example to enter the expression.
2. From the Compute menu, choose Polynomials, and then choose Sort. The computational engine orders the terms and displays this result:

$$3x - 7x^2 + 8x - 3 + x^5 = x^5 - 7x^2 + 11x - 3$$

Write $\frac{x^6 - 5x^4 + 3x + 4}{(x^2 - 2)(x + 3)^3}$ as the sum of partial fractions

1. Click [fraction button].
2. Type **x**, click [Nˣ], type **6**, and press the spacebar.
3. Type **-5x**, click [Nˣ], type **4**, and press the spacebar.
4. Type **+3x+4** and press TAB.
5. Click [(□)], type **x**, click [Nˣ], type **2**, and press the spacebar.

6. Type **-2** and press the spacebar.

7. Click (□), type **x+3**, and press the spacebar.

8. Click Nˣ and type **3**.

9. From the Compute menu, choose Calculus, and then choose Partial Fractions. The computational engine computes the partial fraction and returns this result:

$$\frac{x^6-5x^4+3x+4}{(x^2-2)(x+3)^3} = x - 9 + \frac{319}{7(x+3)^3} - \frac{4491}{49(x+3)^2} + \frac{17\,126}{343(x+3)} + \frac{1}{343}\frac{-534+367x}{x^2-2}$$

Find the roots of the expression x^2+4

1. Click T or, from the Insert menu, choose Math to start mathematics.

2. Type **x**, click Nˣ, type **2**, press the spacebar, and type **+4**.

3. From the Compute menu, choose Polynomials, and then choose Roots. Your screen shows this result:

$$x^2+4, \text{ roots: } \begin{matrix} 2i \\ -2i \end{matrix}$$

Try these exercises with polynomials

Collect the terms in these expressions:

$$5t^2+2t-16t^5+t^3-2t^2+9$$

$$x^2+y+5-3x^3y+5x^2+4y^3+13+2x^4 \text{ (Use the variable } x\text{)}$$

Order the terms in these expressions:

$$3x-7x^2+8x-3+x^5$$

$$x^2+y+5-3x^3y+5x^2+4y^3+13+2x^4 \text{ (Use the variable } x\text{)}$$

Divide these polynomials:

$$\frac{x^5-5x^4+3x+4}{x^2-2} \qquad \frac{(3x^2+3x)}{(8x^2+7)}$$

Find the roots of these expressions:

$$x^3-2x-2x^2+4 \qquad x^3-\frac{13}{5}ix^2-8x^2+\frac{29}{5}ix+\frac{81}{5}x+6i-\frac{18}{5}$$

Compute in Place

You can use many computational commands to compute in place, replacing part of an expression with the result of an operation. You can use this technique to demonstrate the steps involved in a problem solution.

Simplify $\frac{x+2}{x+1} + \frac{3x}{x-1}$

1. Click [fraction button], type **x+2**, press TAB, type **x+1**, and press the spacebar.
2. Type + , click [fraction button], type **3x**, press TAB, and type **x-1**.
3. Click [Evaluate button]. The computational engine returns this result:

$$\frac{x+2}{x+1} + \frac{3x}{x-1} = 2\frac{2x^2+2x-1}{(x+1)(x-1)}$$

4. Select the denominator in the answer, hold down the CTRL key, and click [Evaluate button]. On your screen, you see this result:

$$\frac{x+2}{x+1} + \frac{3x}{x-1} = 2\frac{2x^2+2x-1}{x^2-1}$$

Simplify $\frac{x+2}{x+1} + \frac{3x}{x-1}$ and show the steps in the solution

1. Enter and simplify the expression, follow steps 1–3 in the previous example.
2. Type = .
3. Select the answer returned by the computational engine.
4. Hold down the CTRL key while you drag a copy of the answer to the right of the equals sign.
5. Select the denominator in the copied answer, hold down the CTRL key, and click [Evaluate button]. Now you see all the steps in the solution:

$$\frac{x+2}{x+1} + \frac{3x}{x-1} = 2\frac{2x^2+2x-1}{(x+1)(x-1)} = 2\frac{2x^2+2x-1}{x^2-1}$$

Compute in place to show the steps in the solutions of these problems

Combine $\frac{x}{x-y} - \frac{x^2}{x^2-y^2}$ to give $\frac{xy}{(x-y)(x+y)}$.

Combine these fractions $\frac{x}{x+3} - \frac{6x+6}{x^2-9} + \frac{x+1}{x-3}$ and simplify to $\frac{2x+1}{x+3}$.

Create 2-D and 3-D Plots

In *Scientific WorkPlace* and *Scientific Notebook,* you can create 2-dimensional and 3-dimensional plots of polynomials, trigonometric functions, and exponentials.

Plot x^2

1. Click T or, from the Insert menu, choose Math, then type **x**, click N^x, and type **2**.

2. Click or, from the Compute menu, choose Plot 2D and then choose Rectangular. The computational engine plots the expression x^2:

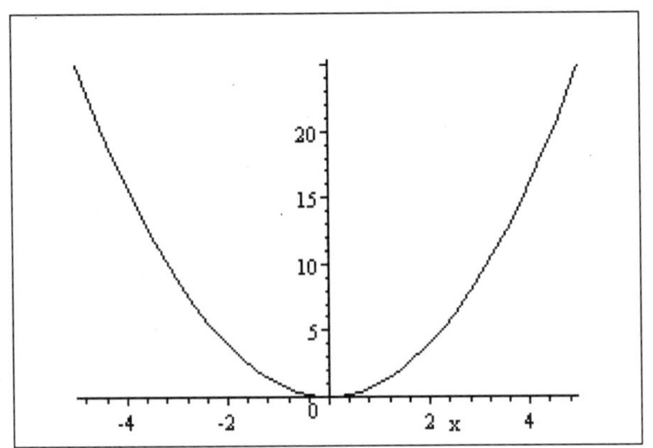

3. Add to the plot:

   a. Press ENTER to start a new line and then start mathematics.

   b. Type **x**, click N^x, and type **3**.

   c. Select the expression x^3 and drag it onto the plot. The computational engine re-plots, adding the new function:

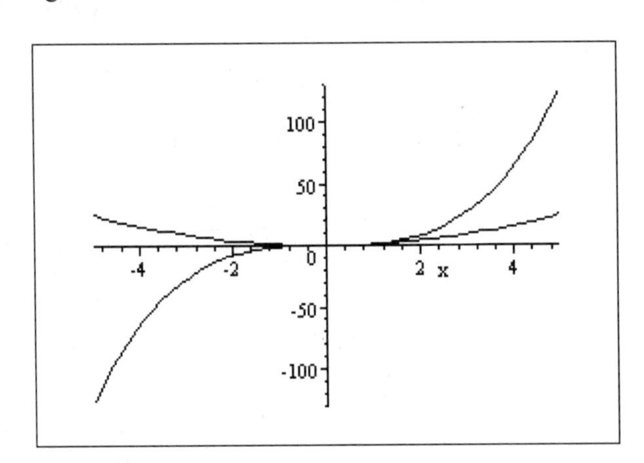

Create a polar plot

1. Click or, from the Insert menu, choose Math, then type **sin2t**.

2. Click or, from the Compute menu, choose Plot 2D and then choose Polar. The computational engine plots the expression $\sin 2t$:

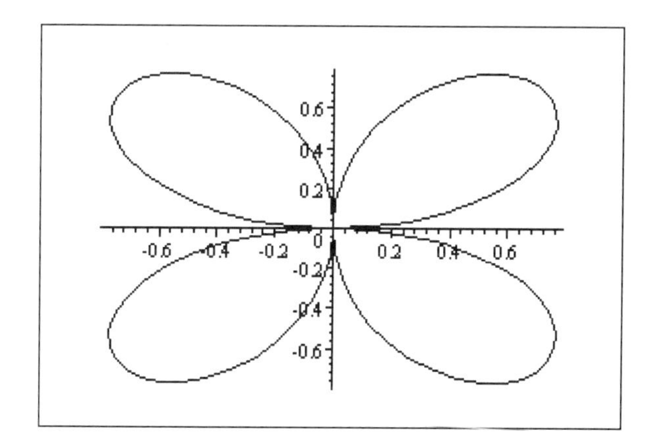

Plot x^2y^3

1. Click or, from the Insert menu, choose Math to start mathematics.

2. Type **x**, click , type **2**, and press the spacebar; then type **y** , click , and type **3**.

3. Click or, from the Compute menu, choose Plot 3D and then choose Rectangular. The computational engine plots the expression x^2y^3:

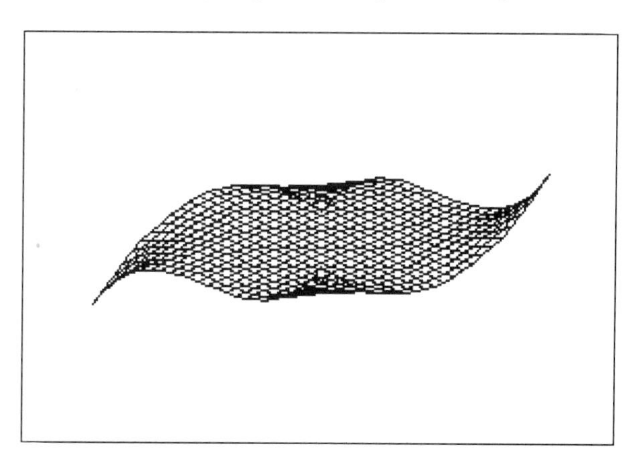

Plot $[x \sin x \cos y, x \cos x \cos y, x \sin y]$

1. Click [□].
2. Type **xsinxcosy,**
3. Type **xcosxcosy,**
4. Type **xsiny**.
5. Click or, from the Compute menu, choose Plot 3D and then choose Rectangular. The computational engine produces a plot whose components you can change.
6. Click and select the Plot Components tab.
7. In the Domain Intervals area, set $0 < x < 6.28$ and set $0 < y < 3.14$.
8. In the Plot Style area, select WireFrame, and choose OK.

   The computational engine replots the expression $[x \sin x \cos y, x \cos x \cos y, x \sin y]$:

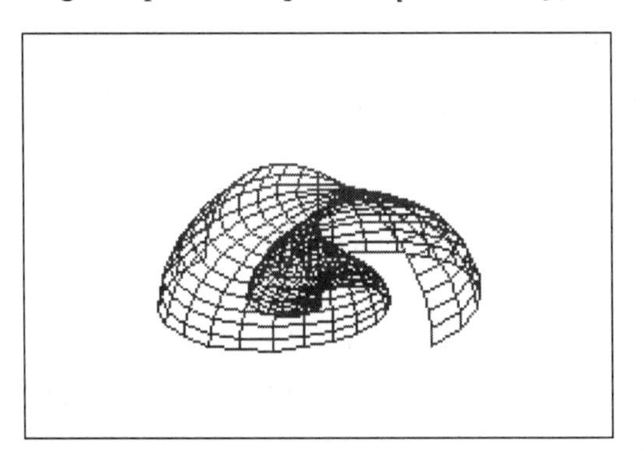

Plot these expressions

Use Plot 2D, Rectangular to plot $x \sin \frac{1}{x}$.

Use Plot 2D, Parametric to plot $(\sin 2t, \cos 3t)$.

Use Plot 2D, Implicit to plot $(x-2)^2 + (y-3)^2 = 25$, where $-3 \le x \le 7$ and $-2 \le y \le 8$.

Use Plot 3D, Cylindrical to plot $1 - z$, where $0 \le \theta \le 2\pi$ and $0 \le z \le 1$.

Use Plot 3D, Spherical to plot 2, where $0 \le \theta \le 2\pi$ and $0 \le z \le 2\pi$.

Perform Matrix Operations

You can perform a full range of matrix operations including addition, multiplication, and inversion.

Transpose the matrix $\begin{bmatrix} 1 & 4 \\ 2 & 5 \\ 3 & 6 \end{bmatrix}$

1. Click [icon], click [icon], and specify 3 rows and 2 columns.
2. Type **1**, press TAB, type **4**, and press TAB.
3. Type **2**, press TAB, type **5**, and press TAB.
4. Type **3**, press TAB, type **6**.
5. From the Compute menu, choose Matrices and then choose Transpose. The computational engine transposes the matrix:

$$\begin{bmatrix} 1 & 4 \\ 2 & 5 \\ 3 & 6 \end{bmatrix}, \text{transpose: } \begin{bmatrix} 1 & 2 & 3 \\ 4 & 5 & 6 \end{bmatrix}$$

Invert the matrix $\begin{bmatrix} 1 & 2 \\ 7 & -3 \end{bmatrix}$

1. Click [icon], click [icon], and specify 2 rows and 2 columns.
2. Type **1**, press TAB, type **2**, press TAB, type **7**, press TAB, and type **-3**.
3. From the Compute menu, choose Matrices and then choose Inverse. The computational engine inverts the matrix:

$$\begin{bmatrix} 1 & 2 \\ 7 & -3 \end{bmatrix}, \text{inverse: } \begin{bmatrix} \frac{3}{17} & \frac{2}{17} \\ \frac{7}{17} & -\frac{1}{17} \end{bmatrix}$$

Perform these matrix operations

Adjugate $\begin{bmatrix} 97 & 50 \\ 56 & 49 \end{bmatrix}$.

Concatenate $\begin{bmatrix} a & 5 \\ 6 & b \end{bmatrix} \begin{bmatrix} 9 & c \\ d & 4 \end{bmatrix}$.

Find the determinant of $\begin{bmatrix} a & b \\ c & d \end{bmatrix}$.

Solve Differential Equations

Many differential equations have answers in terms of familiar functions. You can use the Solve ODE command to find those functions. If you specify initial conditions for an equation, you can solve the equation numerically.

Compute $\frac{dx}{dy} = y$

1. Click .
2. Type **dx**, press TAB, type **dy**, and press the spacebar.
3. Type **=y**.
4. From the Compute menu, choose Solve ODE, and then choose Exact. The computational engine computes the equation and gives this result:

$$\frac{dx}{dy} = y, \text{ Exact solution is: } x(y) = \tfrac{1}{2}y^2 + C_1$$

Find the solution of the initial value problem $\begin{bmatrix} y' = y \\ y(0) = 1 \end{bmatrix}$

1. Click , click on the Math2 toolbar, and specify 2 rows and 1 column.
2. Type **y'=y** and press TAB.
3. Type **y**, click , type **0**, and press the spacebar.
4. Type **=1** and press the spacebar twice.
5. From the Compute menu, choose Solve ODE, and then choose Exact.
6. Type **x** as the independent variable you want the computational engine to use and choose OK. The engine computes the equation and returns this result:

$$\begin{bmatrix} y' = y \\ y(0) = 1 \end{bmatrix}, \text{ Exact solution is: } y(x) = e^x$$

Compute these differential equations

$$\frac{dx}{dy} = 2xy \qquad y'' = y \qquad y'' + y = 0$$

Compute Statistics

Basic and advanced statistical computations are available.

Find the modes of $1,1,3,4,4,4,5,3,8,1,9,5,2$

1. Click [T] or, from the Insert menu, choose Math to start mathematics.
2. Type **1,1,3,4,4,4,5,3,8,1,9,5,2**. Remember to type the commas.
3. From the Compute menu, choose Statistics, and then choose Mode to obtain modes of $1,4$.

Find the medians of the columns of the matrix $\begin{bmatrix} 1.5 & 6.7 \\ 3.9 & 2.2 \\ 5.5 & 4.3 \end{bmatrix}$

1. Click [matrix button], click [matrix grid button], and specify 3 rows and 2 columns.
2. Fill the matrix with the values **1.5**, **6.7**, **3.9**, **2.2**, **5.5**, and **4.3.** Press TAB to move from cell to cell.
3. From the Compute menu, choose Statistics, and then choose Median to obtain medians of $3.9, 4.3$.

Find the standard deviation of $2.5, 6.8, 3.5, 1.9, 2.3, 4.5$

1. Click [T] or, from the Insert menu, choose Math to start mathematics.
2. Enter the numbers **2.5**, **6.8**, **3.5**, **1.9**, **2.3**, and **4.5**, separated by commas.
3. From the Compute menu, choose Statistics, and then choose Standard Deviation. The computational engine computes a standard deviation of 1.8357.

Compute these statistics

Find the mean of these values: $1,2,3,4$ $\begin{bmatrix} 1.5 & 6.7 \\ 3.9 & 2.2 \end{bmatrix}$

Find the variance of $2.5, 6.8, 3.5, 1.9, 2.3, 4.5$.

Using multiple regression, fit a curve to these points (dependent variable in column 2): $\begin{bmatrix} x & y \\ 1 & 2 \\ 2 & 4 \\ 3 & 7 \end{bmatrix}$

Compute with Units of Measure

You can convert values from one unit of physical measure to another and perform computations on equations containing units.

Convert $28.6\,\mathrm{lbf}$ to newtons

1. Click **T** or, from the Insert menu, choose Math to start mathematics.
2. Type **28.6** and then click **ftlb** or, from the Insert menu, choose Unit Name.
3. Select Force, select Pound-force, and then choose Insert.
4. Type **=x** and then in the Unit Name dialog box, select Force, select Newton, and then choose Insert.
5. From the Compute menu, choose Solve, and then choose Exact. This is the result:

$$28.6\,\mathrm{lbf} = x\,\mathrm{N}\text{, Solution is: } \{x = 127.22\}$$

Express $20.0\,\mathrm{mi}\,/\,\mathrm{h}$ in kilometers/minute

1. Click **T** or, from the Insert menu, choose Math to start mathematics.
2. Type **20.0** and then click **ftlb** or, from the Insert menu, choose Unit Name.
3. Select Length, select Mile, and then choose Insert.
4. Type / and then in the Unit Name dialog box, select Time, select Hour, and choose Insert.
5. Type **=x** and then in the Unit Name dialog box, select Length, select Kilometer, and choose Insert.
6. Type / and then in the Unit Name dialog box, select Time, select Minute, and choose Insert.
7. From the Compute menu, choose Solve and then choose Exact. This is the result:

$$20.0\,\mathrm{mi}\,/\,\mathrm{h} = x\,\mathrm{km}\,/\,\mathrm{min}\text{, Solution is: } \{x = .536\,45\}$$

Computations involving units of measure

Express $53.7\,\mathrm{lbf}\,/\,\mathrm{in}^2$ in newtons per square meter.

Compute $3\,\mathrm{A}\;6\,\mathrm{V}$; then convert the answer to watts.

Convert $3.7\,\mathrm{N}\,/\,\mathrm{cm}^2$ to pounds per square inch.

Create Exams and Quizzes

In *Scientific WorkPlace* and *Scientific Notebook,* you can create algorithmically-generated course materials such as exams, quizzes, tests, tutorials, problem sets, drills, and homework assignments. Instead of writing a series of exams, each containing a variation of your questions and answers, you can use the Exam Builder to write a single exam and state the questions and answers with algorithms. Students can drill and practice a skill at length by opening the same Exam Builder file repeatedly, obtaining a slightly different set of questions each time. Similarly, you can test each student with a slightly different version of an exam, with each version drawn from the same Exam Builder source file.

The Exam Builder is appropriate for any level of mathematics instruction—elementary arithmetic, algebra, trigonometry, calculus, linear algebra, differential equations, probability, or statistics. Course materials created with the Exam Builder can be worked and graded online or in the traditional pencil-and-paper form. The online Help system contains detailed instructions for using the Exam Builder.

4 Typesetting Your Document

If you've used other versions of our software, you may find that the biggest change in this version of *Scientific WorkPlace* and *Scientific Word* is the ability to preview and print your documents two ways: either with or without LaTeX typesetting. The two methods produce noticeably different results.

We distinguish the two methods of document production by referring to the processes that involve LaTeX typesetting as *typeset compile, typeset preview,* and *typeset print,* and by referring to the processes that don't involve typesetting as simply *preview* and *print.* You typeset your document using commands on the Typeset menu or the buttons on the Typeset toolbar:

Menu	**Command**	**Button**
Typeset	Compile	
Typeset	Preview	
Typeset	Print	

You produce your document without typesetting using the commands on the File menu or buttons on the Standard toolbar:

Menu	**Command**	**Button**
File	Preview	
File	Print	

Typesetting isn't available in *Scientific Notebook.*

Each time you preview or print your *Scientific WorkPlace* or *Scientific Word* document, you can choose whether or not to typeset it. If you typeset, the program uses LaTeX to compile the document, employing special formatting features such as hyphenation, ligatures, and sophisticated paragraph and line breaking. LaTeX also generates document elements, automatically creating tables of contents, cross-references, citations, bibliographies, and other similar text elements that are called for in the document shell. Therefore, after you have typeset your document, its appearance may be quite different from what you see in the document window.

If you don't typeset your document, the program uses the same routines to preview and print as it does to display the document on the screen. The program doesn't use special formatting features, nor does it automatically generate any document elements. Therefore, the appearance of the previewed or printed document is quite similar to what you see in the document window.

We suggest that you follow these general guidelines:

- When a finely typeset document appearance is a high priority, typeset preview or typeset print the document with the commands on the Typeset menu.
- When a finely typeset document appearance isn't a priority or you need output quickly, preview or print the document with the commands on the File menu.

Understanding LaTeX Typesetting

When you typeset your document, the program previews and prints your documents in several steps. First, your document is compiled with LaTeX to resolve typographic details such as line spacing, page breaks, and cross-references, and to generate any specified document elements. Depending on the complexity of your document, it may be passed through LaTeX more than once. The compilation creates a *device-independent,* or *DVI,* file that contains your typeset document in a form independent of any output device. (Your original document is unchanged.) *Scientific WorkPlace* or *Scientific Word* is suspended until the DVI file has been created. Second, the DVI file is sent to the TrueTeX previewer or to the printer driver.

Scientific WorkPlace and *Scientific Word* are supplied with TrueTeX, which includes a TeX formatter, a TeX screen previewer, and scalable TrueType fonts. You can access the TeX formatter and screen previewer from the icons in the program group. More information about using other previewers and print drivers is available in the online Help.

Typeset Previewing and Typeset Printing

You can typeset preview your document to examine its typeset appearance before you send it to the printer. When you are satisfied with the document, typeset print it, either from the TrueTeX preview window or the document window.

► **To typeset preview a document**

1. Save the document.
2. On the Typeset toolbar, click the Typeset Preview button or, from the Typeset menu, choose Preview.

   The program compiles your document with LaTeX, and then opens the TrueTeX preview screen and displays your document as it will appear in print.
3. Use the scroll bars and the menu commands to move around in the previewer and examine your document.
4. When you're ready to leave the previewer, choose Exit from the File menu.

You can typeset print a document from the TrueTeX preview window or from the document window.

► To typeset print a document from the TrueTEX preview window

1. Typeset preview the document.
2. From the File menu in the TrueTEX preview window, choose Print.
3. Select the printer and print specifications you want, and then choose OK.

► To cancel typeset printing from the TrueTEX preview window

- Click the print status bar at the top of the document image.

► To typeset print a document from the document window

1. Save the document.
2. On the Typeset toolbar, click the Typeset Print button or, from the Typeset menu, choose Print.
3. Make any necessary selections in the Print dialog box, and then choose OK.

If your document doesn't have a current DVI file, the program automatically compiles the document when you typeset preview or typeset print it. Also, you can use the Compile command on the Typeset menu to compile your document independently and then preview or print the typeset document at a different time. The Compile command is active only when you've saved the document and made no further changes.

► To compile a document

1. Save the document.
2. On the Typeset toolbar, click the Typeset Compile button or, from the Typeset menu, choose Compile.
3. Select the options you want from the Compile dialog box, and then choose OK.

Understanding the Appearance of Typeset Documents

When you typeset, the program formats your document using these specifications:

- The *typesetting specifications* for the document shell. The specifications are a collection of LATEX formatting instructions that govern all aspects of the typeset appearance of your document: type face, type size, margins, page size, line spacing, location and appearance of headers and footers, paragraph layout and indentation, section headings, page breaks, and countless other typographic details. The specifications are contained in files with extensions of `.sty`, `.clo`, and `.cls`.
- Any specified LATEX *class options* or LATEX *packages* specified for the document or its shell. Class options and packages contain instructions that modify the typesetting specifications in some way.
- Any additional LATEX commands you have entered in the document itself.

Note The typesetting specifications are set initially by the shell you use to create your document. Some modifications to the shell and the document are possible by adding LaTeX class options and packages and by using commands on the **Typeset** menu. However, we strongly discourage attempts to modify the specifications, particularly if you aren't extremely familiar with TeX and LaTeX. If the shell for your document doesn't produce the typeset results you want, start a new document with a shell that meets your requirements more closely.

The program uses the typesetting specifications *only when you typeset.* When you produce the document without typesetting it, the program ignores the typesetting specifications and instead uses the margins, headers, footnotes, and page numbers specified in the **Page Setup** dialog box and the tag properties specified in the style. Modifications to tag properties that you make from the **Tag Appearance** dialog box *do not affect* the typeset appearance of your document.

In other words, the way you choose to produce your document determines which set of specifications the program uses and, consequently, how your document appears on the preview screen and in print.

When the document is	SW formats according to
Displayed in the document window	Style
Produced without typesetting (From **File** menu, choose **Preview** or **Print**)	Style Page setup specifications Print options
Produced with typesetting (From **Typeset** menu, choose **Compile**, **Preview** or **Print**)	Typesetting specifications LaTeX options and packages LaTeX commands

Creating Typeset Document Elements

Many document shells contain predefined *fields* that instruct LaTeX to generate document elements—especially those in the front matter, such as title pages and abstracts—when you typeset your document. If you don't typeset, the fields are ignored and the elements aren't generated. As you build your document, you can add other fields that automatically create notes, citations, index entries, and cross-references to numbered parts of the document whenever you typeset with LaTeX. Also, if you are very familiar with TeX and know the TeX command for an object or operation not available in *SW,* you can enter it in your *SW* document in a TeX field. The commands to create fields appear on the **Field** menu. Many also appear on the Field and Typeset Field toolbars:

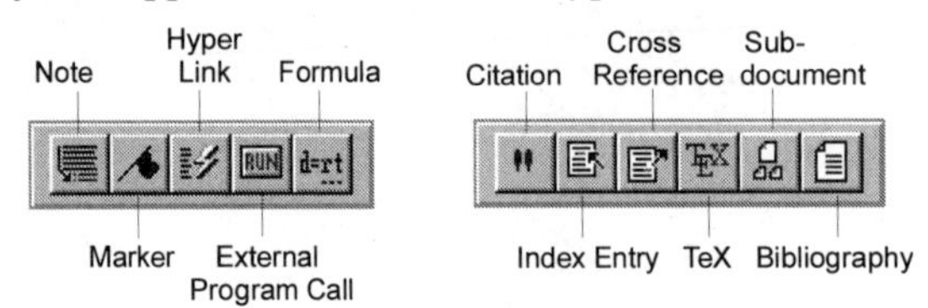

Important Remember that you must typeset your document to generate document elements that are specified in fields. If you produce your document without typesetting it, the program doesn't process the document with LaTeX and the elements aren't generated, so the field appears as a small gray box, just as it does in the document window.

Creating Cross-References

When you typeset your document, you can generate automatic cross-references to other numbered items in your document, such as equations, theorems, graphics, or sections, or to the pages on which they appear. Cross-references have two parts, the reference itself and a *marker* containing a *key* for the numbered item. When you typeset the document, LaTeX resolves the cross-reference, replacing it with the number of the marked item. Note that creating markers for graphics, numbered equations, and manual bibliography items requires a process different from the one described below; see online Help for more information.

► To create a cross-reference to a numbered item

1. Place the insertion point where you want the reference to appear.
2. On the Typeset Field toolbar, click the Cross Reference button or, from the Insert menu, choose Field, and then choose Cross Reference.
3. In the Print area, check Object Counter.
4. Enter the key of the marked item and choose OK.

   The program displays the reference on the screen in a gray box containing the word *ref* and the key you entered, as in this example: ref: main idea .

► To create a marker for a numbered item

1. Place the insertion point where you want the marker.
2. On the Field toolbar, click the Marker button or, from the Insert menu, choose Field, and then choose Marker.
3. In the Key box, enter a unique key for the item, and then choose OK.

   When the Marker Fields display is turned on in the View menu, a marker looks like this in the document window: marker: main idea .
4. Save the document.

Creating Notes

Your documents can contain references to margin notes and footnotes that appear elsewhere in your document. When you typeset, LaTeX generates a cross-reference to the note and formats the note according to the typesetting specifications. If you don't typeset, notes appear in print as they do in the document window; that is, as small gray boxes.

► **To enter a footnote**

1. Place the insertion point where you want the reference to the footnote to appear.
2. On the Field toolbar, click the Note button or, from the Insert menu, choose Field, and then choose Note.
3. In the Note dialog box, select footnote as the type of note.
4. Enter the text of the footnote, and then choose OK.

Creating Bibliographies and Citations

If the document shell supports bibliographies, you can create a bibliography list and automatic cross-references, or *citations,* to items in the list. When you typeset your document, LaTeX generates the citation and formats it correctly, according to the typesetting specifications for the shell.

You can create bibliographies in two ways. If your bibliography is complex, you can create an automatic bibliography using BIBTeX, a public domain program created by Oren Patashnik. A BIBTeX bibliography is convenient if you have a long list of references that you plan to use in other articles or books, because you don't have to create the bibliography list yourself. BIBTeX generates and formats the list automatically by extracting references from a database using the citations you insert into your document. The creation and use of BIBTeX bibliographies is somewhat complex. Please refer to the online Help for detailed instructions.

If your bibliography is a simple one, you can create it manually. A manual bibliography, which works just like a series of cross-references, is convenient when your list of references is short and you don't plan to use those references in other articles or books. You must format entries in a manual bibliography yourself.

► **To specify a manual bibliography**

1. From the Typeset menu, choose Bibliography Choice.
2. Select Manual Entry and choose OK.

▶ **To create a list of bibliography items**

1. Move the insertion point to the end of the line that is to precede the bibliography, and press ENTER.
2. Apply the Bibliography item tag.
3. In the Bibliography Item Properties dialog box, enter a unique key for the item and choose OK.
4. Type the bibliographic information for the item.
5. If you want to create another bibliographic item, press ENTER and then repeat steps 3–4.
6. When you have completed all the entries, press ENTER and then click [<=•] to leave the list.

▶ **To create a citation for an item in a manual bibliography**

1. Place the insertion point where you want the citation to appear.
2. On the Typeset Field toolbar, click the Citation button [■] or, from the Insert menu, choose Field, and then choose Citation.
3. In the Citation dialog box, enter the key for the bibliography item you want to cite, and then choose OK.

In the document window, you see the citation in a small gray box containing the key for the bibliography item. When you typeset the document, LaTeX uses the number of the item in place of the key.

Obtaining More Information About Typesetting

A detailed discussion of LaTeX typesetting is beyond the scope of this booklet. However, you can find much more information about typesetting with *Scientific WorkPlace* and *Scientific Word* in the online Help and in *Creating Documents with Scientific Word and Scientific WorkPlace, Version 3.5*. You can find more information about LaTeX in *LaTeX, A Document Preparation System* by Leslie Lamport and in *The LaTeX Companion* by Michel Goossens, Frank Mittelbach, and Alexander Samarin. Also, you can find information about TeX in *The TeXbook* by Donald E. Knuth.

5 Getting the Most from Your Software

Your *SW* software is straightforward and easy to use, especially if you take advantage of the many resources available to you. *Scientific WorkPlace, Scientific Word,* and *Scientific Notebook* are supplied with documentation including a Help system and tutorial materials that will help you learn how to create documents and enter mathematics with *SW*. If you have an Internet connection, additional documentation—in particular, the Help system for the *Scientific Notebook Viewer*—is available from the *SW* web site.

Additionally, two manuals provide program details: *Creating Documents with Scientific WorkPlace and Scientific Word, Version 3.5* provides information about entering and editing mathematics; formatting documents; structuring documents for typesetting and for use online; maintaining files; and customizing your installation. Much of the information in the manual contains is applicable to *Scientific Notebook*. *Doing Mathematics with Scientific WorkPlace and Scientific Notebook* explains how to use the built-in computer algebra systems Maple and MuPAD to do mathematics without dealing directly with the syntax of the computer algebra system. Along with examples and exercises, the manual provides basic procedures for using the system to compute and plot and to solve problems in analytic geometry, calculus, linear algebra, vector analysis, differential equations, statistics, and applied modern algebra.

If you can't find the information you need in one of these resources, technical support is available. We also regularly make additional information available on our unmoderated discussion forum and electronic mail list.

Using Online Help

SW includes an extensive Help system available online or from your CD-ROM. While you're working in *SW*, you can get information quickly from the online Help feature. You can search Help to find basic and advanced information about *SW* commands and operations, including instructions for using the built-in computational power of *SW* to perform numeric, symbolic, and graphic computations. If you save copies of the Help documents in *Scientific WorkPlace* or *Scientific Notebook*, you can interact with the mathematics they contain, experimenting with or reworking the included examples. In addition, two *SW* programs—the Style Editor and the Document Manager—have their own online Help.

► **To use the Help system**

Choose	To
Contents	See a list of online information
Search	Find a Help topic
Index	Access the online index to General Information, Computing Techniques, or the Reference Library
Resource Center	Open the link to the *Scientific Notebook* resource center
Scientific Notebook Web Site	Open the link to the *Scientific Notebook* web site
MacKichan Software Web Site	Open the link to the MacKichan Software, Inc. web site
Register	Register your software and obtain a license
System Features	See a list of available features; change the serial number for your installation
License Information	Obtain information about registering your system
About SW	Obtain information about your *SW* installation

Also, we urge you to explore the supplemental technical documents supplied with the program. You can use *SW* to open, view, and print the documents. In particular, we suggest you read the following documents:

- In the `Help\general` directory, the document `35techref.tex`, which contains technical information on the features in Version 3.5.
- The documents in the `Play` directory, which demonstrate the use of computation in *Scientific WorkPlace* and *Scientific Notebook.*
- The documents in the `SNSamples` directory, which are examples of online documents created with *SW.*
- The documents in the `SWSamples` directory, which are examples of documents intended for typesetting in *Scientific WorkPlace* and *Scientific Word.*

Obtaining Technical Support

If you can't find the answer to your questions in the manuals or the online Help, you can obtain technical support from our Web-based Technical Support forum at

http://www.mackichan.com/support.html

You can also contact our Technical Support staff by electronic mail, telephone, or fax. We urge you to submit questions by electronic mail whenever possible in case our technical staff needs to obtain your file to diagnose and solve the problem.

When you contact us by electronic mail or fax, please provide complete information about the problem you're trying to solve. We must be able to reproduce the problem exactly from your instructions. When you contact us by telephone, you should be sitting at your computer with *SW* running.

Please provide the following information any time you contact Technical Support:

- The *SW* product you have installed.
- The version number of your *SW* installation.
- The serial number of your *SW* installation.
- The version of the Windows system you're using.
- The type of hardware you're using, including network hardware.
- A description of what happened and what you were doing when the problem occurred.
- The exact wording of any messages that appeared on your computer screen.

► To contact technical support

- Contact Technical Support by electronic mail, fax, or telephone between 8 a.m. and 5 p.m. Pacific Time:

Internet electronic mail address: support@mackichan.com
Fax number: (206) 780-2857
Telephone number: (206) 780-2799

Obtaining Additional Information

You can learn more about *SW* on our web site, which is updated regularly to provide the latest technical information about the software. The site also houses links to other TeX and LaTeX resources. We maintain an unmoderated discussion forum and an unmoderated electronic mail list so our users can share information, discuss common problems, and contribute technical tips and solutions. You can link to these valuable resources from our home page.

► To visit our home page

- Use your web browser to access our home page at

http://www.mackichan.com

Learning SW

You can learn a great deal about *SW* just by working with it. Start by opening the program, typing a few sentences, entering some mathematics, and then previewing and printing your document, as described in Chapter 1 "Tools for Scientific Creativity." We also urge you to work through the tutorial exercises, available from the Help Contents. The exercises guide you step-by-step as you create several increasingly complex documents and learn how to enter a variety of mathematical expressions, compute while working in your document, and print with and without typesetting.

▶ **To open the tutorial exercises**

1. From the Help menu, choose Contents.
2. Choose Learn the Basics.
3. Choose the exercise you want:
   - "Before You Start" provides information about the notation, terminology, and instructions used in the exercises.
   - "Creating a Simple Document" provides instructions for creating a basic document.
   - "Printing and Typesetting" explains two ways to produce your document.
   - "Creating an Advanced Document" gives instructions for creating a somewhat more complex document in *Scientific WorkPlace* or *Scientific Word.* The exercise, which takes about 90 minutes to complete, focuses on document structuring, guiding you step-by-step through the unique procedures for creating titles, headings, and theorem statements. Because the document you create contains equations, the exercise also illustrates the straightforward entry and editing of mathematics in *SW.*
   - "Creating Mathematics" presents a series of mathematical examples. The exercises give step-by-step instructions for entering a variety of mathematical expressions using the mouse and via the keyboard.
   - "Performing Computations" presents a series of mathematical computation exercises. The step-by-step instructions guide you through basic computational procedures for *Scientific WorkPlace* and *Scientific Notebook.*

A Toolbars and Buttons

Common Symbols Toolbar

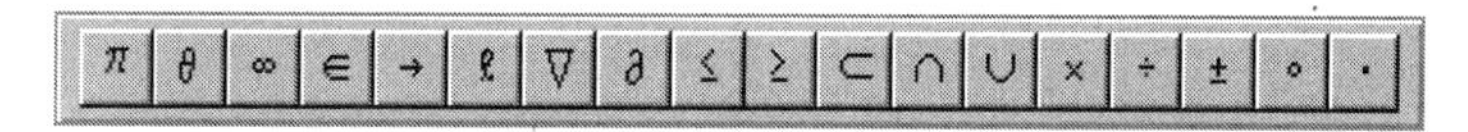

Compute Toolbar

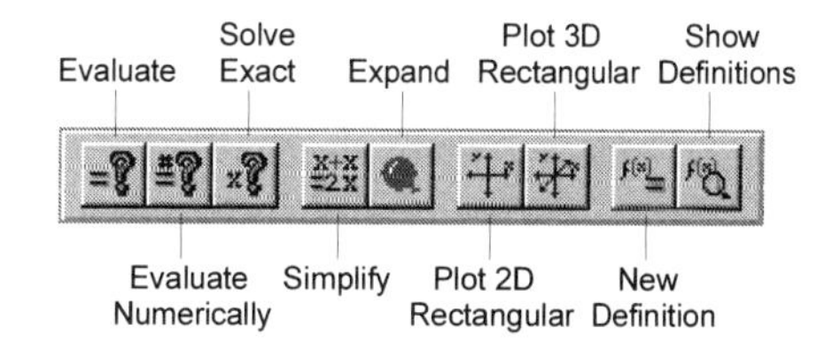

Field Toolbar

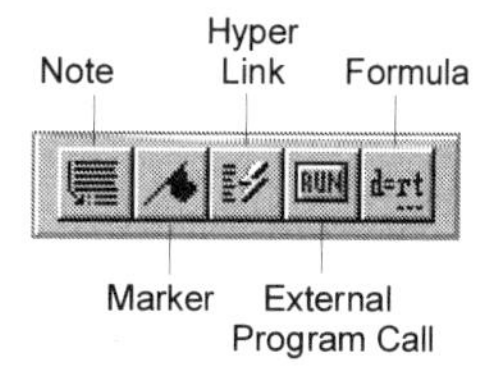

Fragments Toolbar

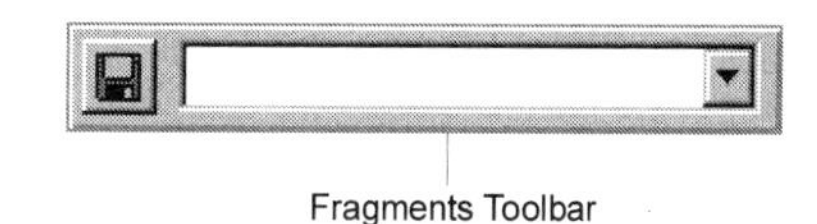

History Toolbar

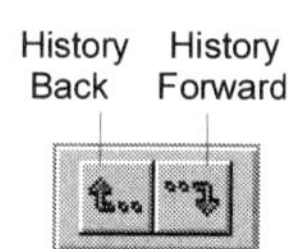

Link Toolbar

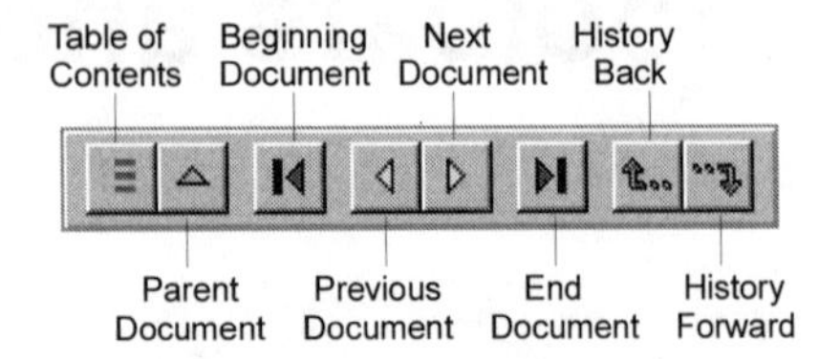

Math Toolbars

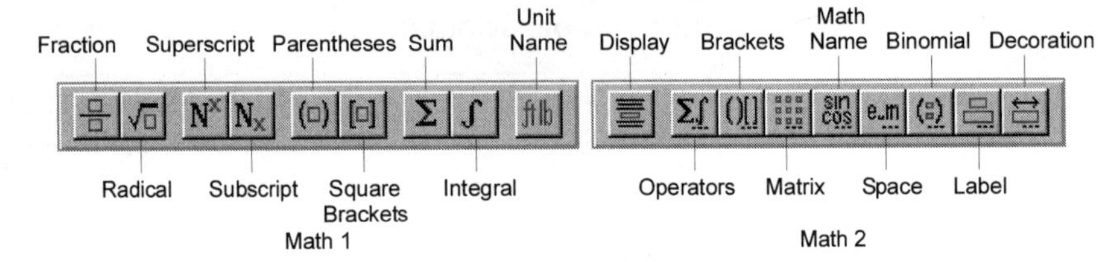

Navigate Toolbar

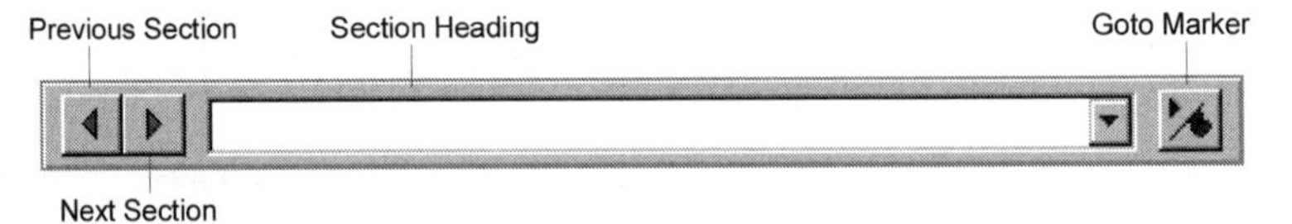

Standard Toolbar

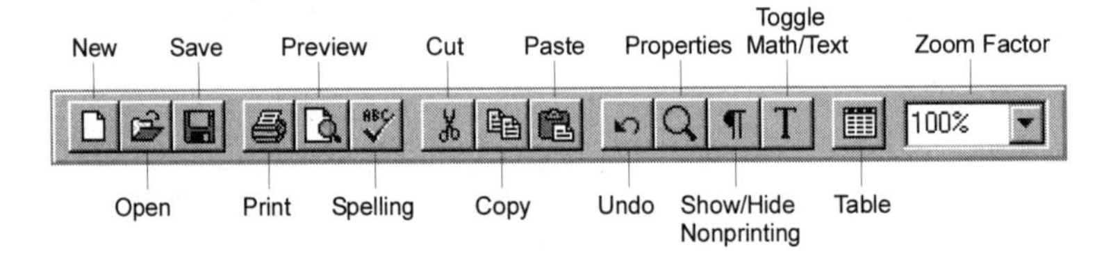

Stop Toolbar

Symbol Toolbar

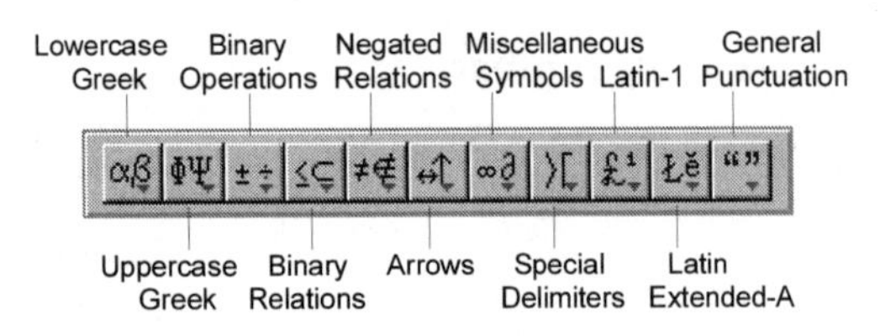

Tag Toolbar

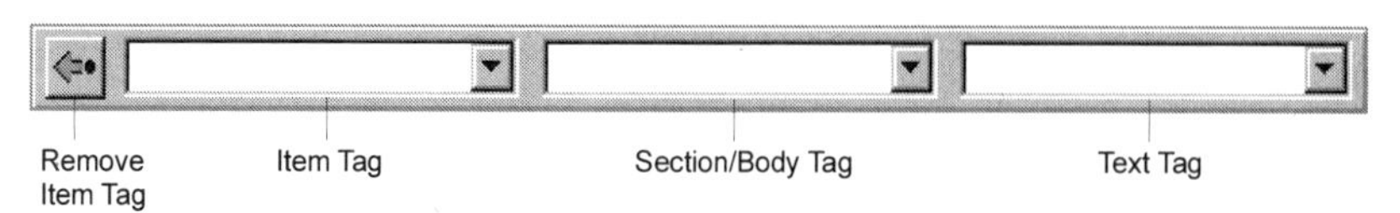

Typeset Toolbar

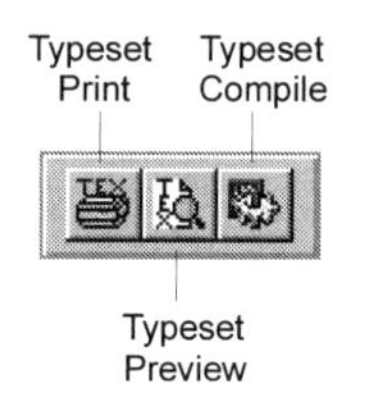

Typeset Field Toolbar

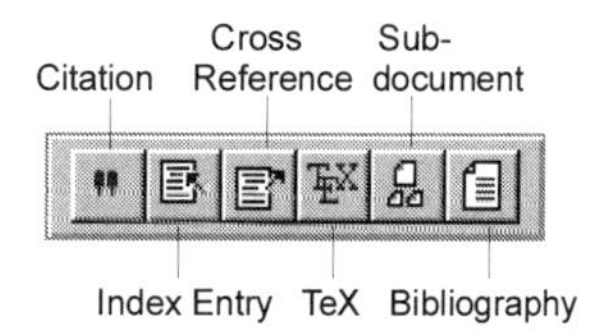

B Keyboard Shortcuts

Scrolling and Editing

Scrolling

To move	Press
To the left	LEFT ARROW
To the right	RIGHT ARROW
Up	UP ARROW
Down	DOWN ARROW
To start of the line	HOME
To end of the line	END
To next screen	PAGE DOWN
To previous screen	PAGE UP
To document start	CTRL+HOME
To document end	CTRL+END
To next field inside a template	TAB or ARROW KEYS
To previous field inside a template	SHIFT+TAB or ARROW KEYS
To outside a template	RIGHT ARROW or LEFT ARROW or SPACEBAR
To the word to the right of the insertion point	CTRL+RIGHT ARROW
To the word to the left of the insertion point	CTRL+LEFT ARROW
Between open documents	CTRL+TAB

Editing

To	Press
Copy the selection to clipboard	CTRL+C
Cut the selection to clipboard	CTRL+X
Paste from clipboard	CTRL+V
Edit Properties	CTRL+F5
Undo the last deletion	CTRL+Z

Selecting

To select	Press
The following screen	SHIFT+PAGE DOWN
The previous screen	SHIFT+PAGE UP
The word to the right of the insertion point	CTRL+SHIFT+RIGHT ARROW
The word to the left of the insertion point	CTRL+SHIFT+LEFT ARROW
The object or symbol to the left of the insertion point	SHIFT+LEFT ARROW
The object or symbol to the right of the insertion point	SHIFT+RIGHT ARROW
Everything between the insertion point and the start of the line	SHIFT+HOME
Everything between the insertion point and the end of the line	SHIFT+END
Everything between the insertion point and the start of the document	CTRL+SHIFT+HOME
Everything between the insertion point and the end of the document	CTRL+SHIFT+END
To choose a command	ALT+the Accelerator keys (the underlined letters for the menu and command)

Entering Mathematics and Text

Toggling Between Mathematics and Text

To	Press
Toggle math/text	CTRL+M or CTRL+T

Entering Mathematical Accents

To enter accents	Press
$\hat{a}$	CTRL+^(CTRL+SHIFT+6)
$\tilde{a}$	CTRL+~(CTRL+SHIFT+‘)
$\acute{a}$	CTRL+’
$\grave{a}$	CTRL+‘
$\dot{a}$	CTRL+.
$\ddot{a}$	CTRL+” (CTRL+SHIFT+’)
$\bar{a}$	CTRL+=
$\vec{a}$	CTRL+–

Entering Mathematical Objects

To enter	Press		
Fraction	CTRL+F	or CTRL+/	or CTRL+1
Radical	CTRL+R	or CTRL+2	
Superscript	CTRL+H	or CTRL+3	or CTRL+UP ARROW
Subscript	CTRL+L	or CTRL+4	or CTRL+DOWN ARROW
Summation	CTRL+7		
Integral	CTRL+I	or CTRL+8	
Brackets	CTRL+9	or CTRL+0 or CTRL+)	or CTRL+(or CTRL+5
Square brackets	CTRL+[	or CTRL+]	or CTRL+6
Braces	CTRL+{	or CTRL+}	
Display	CTRL+D		
2 x 2 Matrix	CTRL+A		
Product	CTRL+P		
Absolute value	CTRL+|		
Thin space	CTRL+SPACEBAR		
Thick space	SHIFT+SPACEBAR		
Em space	CTRL+SHIFT+SPACEBAR		
“ (double open quote)	Single open quote (‘) twice		
- (intraword dash or hyphen)	Hyphen (-)		
– (en dash)	Hyphen (-) twice		
— (em dash)	Hyphen (-) three times		
- (Discretionary hyphen)	CTRL+– (CTRL + hyphen two times)		
¿ (inverted question mark)	? followed by ‘		
¡ (inverted exclamation point)	! followed by ‘		

Entering Symbols and Characters

To enter	Press CTRL+S then press	To enter	Press CTRL+S then press
$\rightarrow$	1	$\subset$	c
$\uparrow$	2	$\vee$	v
$\leftarrow$	3	$\bullet$	b
$\downarrow$	4	∇	n
$\supseteq$	5	$\Downarrow$	$\$$
$\cap$	6	$\Rightarrow$	!
$\subseteq$	7	$\Uparrow$	@
$\cup$	8	$\Leftarrow$	#
$(\square)$	9 or 0 or (or)	$\supset$	%
$\equiv$	-	$\cong$	_
$\neq$	=	$\pm$	+
$\approx$	w	$\aleph$	W
$\in$	e	$\notin$	E
$\sqrt{\square}$	r or R	∞	I
$\otimes$	t or T	$\wp$	P
$\int$	i	$\{\square\}$	{ or }
$\emptyset$	o	$\forall$	A
$\prod$	p	$\oplus$	S
$[\square]$	[or]	$\diamond$	D
$\angle$	a	$\div$	X
$\sum$	s	$\cdot$	C
∂	d	$\wedge$	V
$\frac{\square}{\square}$	f or F	$\neg$	N
$a^{\square}$	h or H	$\leq$	<
$a_{\square}$	l or L	$\geq$	>
$\times$	x	$\exists$	z
$\begin{array}{\|c\|c\|}\hline \square & \square \\ \hline \square & \square \\ \hline \end{array}$	m		

Entering Greek Characters

To enter		Press CTRL+G then press	To enter		Press CTRL+G then press
alpha	α	a	pi	π	p
beta	β	b		Π	P
gamma	γ	g		ϖ	v
	Γ	G	rho	ρ	r
delta	δ	d		ϱ	R
	Δ	D	sigma	σ	s
epsilon	ε	e		Σ	S
	ϵ	E		ς	T
zeta	ζ	z	tau	τ	t
eta	η	h	upsilon	υ	u
theta	θ	y		Υ	U
	ϑ	Z	phi	ϕ	f
	Θ	Y		Φ	F
iota	ι	i		φ	j
kappa	κ	k	chi	χ	q
	$\varkappa$	K	psi	ψ	c
lambda	λ	l		Ψ	C
	Λ	L	omega	ω	w
mu	μ	m		Ω	W
nu	ν	n	digamma	$\digamma$	I
xi	ξ	x			
	Ξ	X			

Entering ANSI Characters

The ANSI codes depend on the Windows code page in use. The sequences in the table below are for U.S. Windows systems.

▶ To enter an ANSI character

1. Hold down the ALT key.
2. On the numeric keypad, type **0** and the number for the ANSI character you want.
3. Release the ALT key.

To enter	Type 0 +	To enter	Type 0 +	To enter	Type 0 +
space	160	À	192	à	224
¡	161	Á	193	á	225
¢	162	Â	194	â	226
£	163	Ã	195	ã	227
¤	164	Ä	196	ä	228
¥	165	Å	197	å	229
¦	166	Æ	198	æ	230
§	167	Ç	199	ç	231
¨	168	È	200	è	232
©	169	É	201	é	233
ª	170	Ê	202	ê	234
«	171	Ë	203	ë	235
¬	172	Ì	204	ì	236
–	173	Í	205	í	237
®	174	Î	206	î	238
¯	175	Ï	207	ï	239
°	176	Ð	208	ð	240
±	177	Ñ	209	ñ	241
^{2}	178	Ò	210	ò	242
^{3}	179	Ó	211	ó	243
´	180	Ô	212	ô	244
µ	181	Õ	213	õ	245
¶	182	Ö	214	ö	246
·	183	×	215	÷	247
¸	184	Ø	216	ø	248
^{1}	185	Ù	217	ù	249
º	186	Ú	218	ú	250
»	187	Û	219	û	251
$\frac{1}{4}$	188	Ü	220	ü	252
$\frac{1}{2}$	189	Ý	221	ý	253
$\frac{3}{4}$	190	Þ	222	þ	254
¿	191	ß	223	ÿ	255

Index

Additional Software

MathTalk/Scientific Notebook®

MathTalk/Scientific Notebook, created by Metroplex Voice Computing, provides voice input for *Scientific Notebook*, enabling you to enter even the most complex mathematics with voice commands. You can use it in conjunction with the keyboard and mouse to speed the entry of text and mathematics, or to completely replace the keyboard and mouse. Used with the MAVIS braille output for *Scientific Notebook*, it provides dramatic new power to visually impaired individuals. *MathTalk/Scientific Notebook* requires Dragon Dictate, which is not included.

Books

Doing Mathematics with Scientific WorkPlace® and Scientific Notebook® Version 3.5

By Darel W. Hardy and Carol L. Walker

Doing Mathematics with Scientific WorkPlace and Scientific Notebook describes how to use the built-in computer algebra systems, Maple® and MuPAD™, to do a wide range of mathematics without dealing direclty with the syntax of the computer algebra system. This book can also be used for version 3.0. Where there are differences in the behavior of the two versions, both behaviors are described.

Creating Documents with Scientific WorkPlace® and Scientific Word® Version 3.5

By Roger Hunter and Susan Bagby

Creating Documents with Scientific WorkPlace and Scientific Word will give you an overview of the process of creating beautiful documents using these two powerful software programs. It covers basic editing and entering of mathematical expressions as well as tables, graphics, lists, indexes, cross-references, tables of contents, and large document management. If you are using *Scientific WorkPlace* or *Scientific Word* to prepare documents for publication, this book is highly recommended.

(Continued)

Books (cont.)

Doing Calculus with Scientific Notebook®

By Darel W. Hardy and Carol L. Walker

Take the mystery out of doing calculus with this must-have companion to the *Scientific Notebook* software. This book provides activities to work with *Scientific Notebook* that will help develop a clearer understanding of calculus. Think of *Scientific Notebook* as a laboratory for mathematical experimentation and *Doing Calculus with Scientific Notebook* as a lab manual of experiments to perform.

An Interactive Introduction to Mathematical Analysis

By Jonathan Lewin (Cambridge University Press)

This book is a sequel to *An Introduction to Mathematical Analysis.* It includes an on-screen hypertext version for reading with *Scientific Notebook.* This on-screen version contains alternative approaches to material, more fully explained forms of proofs of theorems, sound movie versions of proofs of theorems, interactive exploration of mathematical concepts using the computing features of *Scientific Notebook*, automatic links to the author's website for solutions to exercises, and more. It will be available in 2000. Inspection copies of the on-screen version can be requested from the author at lewins@mindspring.com.

Pre-Calculus with Scientific Notebook®

By Jonathan Lewin (Kendall Hunt Publishing Company)

This book contains a standard printed version and an on-screen hypertext version designed for interactive reading with *Scientific Notebook.* The on-screen version includes links to solutions to exercises that reside on the author's website. It can be ordered at any bookstore.

Exploring Mathematics with Scientific Notebook®

By Wei-Chi Yang and Jonathan Lewin (Springer-Verlag)

This book is supplied both in printed form and in an on-screen hypertext version for interactive reading with *Scientific Notebook.* It contains a sequence of modules from a variety of mathematical areas and demonstrates in each how the editing, internet, and computing features of *Scientific Notebook* can be combined to deepen the reader's understanding of mathematical concepts. It can be ordered at any bookstore.

TO ORDER: Visit our webstore, fax, email, or phone us.
Website: www.mackichan.com • Fax: 206-780-2857 • Email: info@mackichan.com • Toll Free: 877-SCI-WORD
600 Ericksen Ave. NE • Suite 300 • Bainbridge Island, WA 98110

Workshops

Scientific WorkPlace®, Scientific Word®, and Scientific Notebook®

By Jonathan Lewin

Seminars, workshops, and training sessions by Jonathan Lewin are available at professional conferences and, by arrangement, at individual campuses of high schools, colleges, and universities.

Scientific Notebook: This presentation introduces the editing, internet and computing features of *Scientific Notebook* documents, publication of mathematical material on websites, and how *Scientific Notebook* can be used as an electronic whiteboard in the classroom.

Scientific WorkPlace and Scientific Word: This presentation introduces the editing and typesetting features of *Scientific WorkPlace* and *Scientific Word*. Participants will be trained in the production of documents that will be printed as professional quality hard copy or submitted to an editor for publication.

Each participant will be given a CD containing sound movies that review the material covered in the workshops. Contact Jonathan Lewin for details at lewins@mindspring.com or 770-973-5931.